Advances in Green Infrastructure Planning

Advances in Green Infrastructure Planning

Editor

Hyun-Woo Kim

MDPI • Basel • Beijing • Wuhan • Barcelona • Belgrade • Manchester • Tokyo • Cluj • Tianjin

Editor
Hyun-Woo Kim
Department of Urban Policy and
Administration
Incheon National University
Incheon
Korea, South

Editorial Office
MDPI
St. Alban-Anlage 66
4052 Basel, Switzerland

This is a reprint of articles from the Special Issue published online in the open access journal *Sustainability* (ISSN 2071-1050) (available at: www.mdpi.com/journal/sustainability/special_issues/green_infrastructure_planning).

For citation purposes, cite each article independently as indicated on the article page online and as indicated below:

LastName, A.A.; LastName, B.B.; LastName, C.C. Article Title. *Journal Name* **Year**, *Volume Number*, Page Range.

ISBN 978-3-0365-3712-2 (Hbk)
ISBN 978-3-0365-3711-5 (PDF)

Contents

About the Editor

Hyun-Woo Kim

Dr. Hyun-Woo Kim is Associate Professor in the Department of Urban Policy and Administration at Incheon National University, South Korea. He received his bachelor's degree at Yonsei University majoring in Urban Planning and Engineering, M.A. in Community and Regional Planning from University of Texas at Austin, and Ph.D. in Urban and Regional Sciences from Texas A&M University, USA. His research focuses on urban green infrastructure planning, stormwater management, urban regeneration and hazard management.

Preface to "Advances in Green Infrastructure Planning"

The expansion of urban areas has facilitated the conversion of undeveloped lands, which has led to environmental degradation, such as loss of habitats, hydro-modification, and the collapse of existing ecosystems. Recent climate change has exacerbated these damages by causing more frequent and serious hazards. To attenuate the impacts of urbanization and the negative effects of climate change, green infrastructure (GI) planning (e.g., nature-based strategies, technologies, policies, and solutions) has arisen as an important approach for balancing urban development and nature. GI offers a variety of benefits to our cities by reducing stormwater runoff, heat waves, and air pollution; expanding wildlife habitats; and increasing recreational opportunities and even nearby property values.

While many studies have revealed various positive effects of GI, further contributions to the linkage of hubs and corridors; the adoption of up-to-date, smart, low-impact developments practices; urban environmental and hazard management planning; and indirect physical or social effects are still encouraged to promote more systematic and sustainable decision-making processes for different types of green spaces.

Hyun-Woo Kim
Editor

 *sustainability*

Article

Machine Learning Simulation of Land Cover Impact on Surface Urban Heat Island Surrounding Park Areas

Dakota McCarty [1], Jaekyung Lee [2] and Hyun Woo Kim [3,*]

1 Department of Urban Planning & Policy, Incheon National University, Incheon 22012, Korea; dakota.mccarty@inu.ac.kr
2 Department of Urban Design & Planning, Hongik University, Seoul 04066, Korea; jklee1@hongik.ac.kr
3 Department of Urban Policy & Administration, Incheon National University, Incheon 22012, Korea
* Correspondence: kimhw@inu.ac.kr

Abstract: The urban heat island effect has been studied extensively by many researchers around the world with the process of urbanization coming about as one of the major culprits of the increasing urban land surface temperatures. Over the past 20 years, the city of Dallas, Texas, has consistently been one of the fastest growing cities in the United States and has faced rapid urbanization and great amounts of urban sprawl, leading to an increase in built-up surface area. In this study, we utilize Landsat 8 satellite images, Geographic Information System (GIS) technologies, land use/land cover (LULC) data, and a state-of-the-art methodology combining machine learning algorithms (eXtreme Gradient Boosted models, or XGBoost) and a modern game theoretic-based approach (Shapley Additive exPlanation, or SHAP values) to investigate how different land use/land cover classifications impact the land surface temperature and park cooling effects in the city of Dallas. We conclude that green spaces, residential, and commercial/office spaces have the largest impacts on Land Surface Temperatures (LST) as well as the Park's Cooling Intensity (PCI). Additionally, we have found that the extent and direction of influence of these categories depends heavily on the surrounding area. By using SHAP values we can describe these interactions in greater detail than previous studies. These results will provide an important reference for future urban and park placement planning to minimize the urban heat island effect, especially in sprawling cities.

Keywords: urban heat island; urbanization; shapley additive explanation; park characteristic; extreme gradient boost; Dallas; land use land cover

Citation: McCarty, D.; Lee, J.; Kim, H.W. Machine Learning Simulation of Land Cover Impact on Surface Urban Heat Island Surrounding Park Areas. *Sustainability* **2021**, *13*, 12678. https://doi.org/10.3390/su132212678

Academic Editor: Salvador García-Ayllón Veintimilla

Received: 20 October 2021
Accepted: 15 November 2021
Published: 16 November 2021

Publisher's Note: MDPI stays neutral with regard to jurisdictional claims in published maps and institutional affiliations.

1. Introduction

Green spaces and landscapes, both natural and man-made, provide diverse ecological benefits, such as surface runoff mitigation [1,2], air quality improvement [3], biodiversity increment [4], and heat wave reduction [5] to urban areas. Particularly, the Urban Heat Island (UHI) effect refers to the phenomenon that temperatures of an urban area are higher than the temperatures of the surrounding area [6–8]. The primary cause of this environmental phenomenon has been attributed to the process of urbanization. Only 15% of the world's population lived in urban areas and in 2020 that number was over 50% [9]. As places urbanize, there comes with it a change to the urban land cover, primarily via the action of replacing the natural vegetation cover with built-up areas, which can come in many forms but generally are seen as an increase in commercial, residential, and industrial zoning [10–13]. This process leads to changes in the physical properties of the land, which in turn leads to an increase in land surface temperature. As the materials making up the surface of the environment change, so do the refraction rates, heat capacity, water retention, and multiple other physical factors that in turn increase land surface temperatures (LST) in the newly developed areas [14]. Thus, with this increase in temperatures in the urbanizing areas disproportionately from the areas outside of the city, there arises the problems of urban heat islands. For some time, UHI research has been a hot topic of research globally [15–19], which has led to many breakthroughs and improvements in the field.

This study focuses on creating a more robust and correct methodology that allows us a more in-depth understanding of not what causes UHI but what influences it the most. More specifically, we are developing and testing a methodology using state-of-the-art techniques (XGBoost and SHAP values) to understand which Land Use Land Cover (LULC) category type has the largest influence on Land Surface Temperature (LST). Additionally, we hypothesize, along with other researchers, that green spaces have the largest impact on LST reduction, and thus UHI, leading us to incorporate an additional test on LULC influence on Park Cooling Intensity (PCI).

Globally, cities have grown exponentially. Large cities have become commonplace and have been studied more intensively. However, there has been less focus on urban areas with a large total area but comparatively smaller population densities, typically being cities experiencing large amounts of urban sprawl even though both high- and low-density cities can experience issues with UHI. For example, the Dallas–Fort Worth metropolitan area, between 2001 to 2011, experienced a 0.148 °C increase in urban LST. This increase has been attributed to the rapid urbanization of the area especially in the suburbs, which, as previously mentioned, is a known cause of increased UHI effects. Additionally, UHI is now considered a cause of severe drought in the metropolitan and surrounding areas in 2011 [20]. Recent literature on urban heat island generally indicates that the artificial increase of temperatures in cities is happening largely due to changes in the built-up environment [12,21–23]. The increased temperatures effect multiple layers of the urban microclimate from the ground surface to canopy layer (buildings, city features, and people), as well as from the canopy layer to the boundary layer (approximately 1500 m above the ground surface) [24]. This indicates that the increase in temperatures has an effect larger than just at the street level of cities, as identified at the 2011 North-Texas drought case.

In small-scale temperature studies, the UHI is mainly characterized by the actual measured air temperature [25–27]. These kinds of studies mainly focused on the temperature difference between the green space and other land types in the city, and the method of characterizing UHI intensity. In addition, some other studies have added microclimate factors and Local Climate Zone (LCZ) factors in their research to investigate UHI [7,12,28,29]. These factors include microclimate conditions such as wind speed, wind direction, humidity, solar light intensity, surface reflectance, and other localized effects on temperature [30–32]. Those studies showed that green spaces have a cooling effect on the air primarily due to the transpiration of the plants in the spaces; this then contributes to lower UHI levels. In addition, local wind speed and wind direction have been shown to have the ability to modify air temperature. Researchers have also found that higher surface albedo correlates with lower temperatures [10]. In short, there is an innate complexity in local climates and related environmental conditions.

However, the downside to microclimate data is in its availability and accuracy at a small scale. The measurement of air temperature is limited by the monitoring system, including pieces of equipment, experts, methods, and related factors. The accuracy of data collection is a key factor, and it is difficult to conduct ideal UHI research on a wide range of space-time scales. These data sources are, however, useful in understanding the generalized sources of data studied. While there are methods such as data interpolation and a growing number of ground-based weather monitoring stations, this data area still presents a challenge to researchers.

In Urban Cold Island (UCI) studies, researchers aim to investigate what helps to cool a city. A major area of research in this field is the investigation of parks and green spaces' effects on UHI mitigation [33,34].

There are two main methods used to quantify the cooling effect of parks: Green-space Cooling Intensity (GCI) and Park Cooling Intensity (PCI) [35,36]. GCI is defined as the temperature difference between green space and the average temperature of the whole study area, while the PCI is typically determined as the temperature difference between inside the park area and the urban areas within a specified buffer area [37,38]. While both methods are used to describe the cooling effect of green spaces and parks, for the purpose of this paper and to better understand the impact of parks on decreasing LST in their immediate area, PCI was chosen.

Reiterating, through the utilization of machine learning boosting algorithms, this research aims to determine which LULC classifications have the largest impact on LST and thus UHI; as well as assess the feasibility of applying these new state-of-the-art machine learning models, being XGBoost, and more modern analysis tools, being SHAP values, to this area of research. The greatest benefit of the XGBoost model is its ability to analyze a large range and scale of data without worries of null values; and as the main model is tree-based, there is a significantly decreased risk of overfitting. In addition, it is borrowed from a game theoretic approach known as SHapley Additive exPlanations (SHAP). SHAP can give a more detailed understanding of which factor has the largest impact on the model. It does this by removing each factor one at a time and trying a combination of factors to see how the combinations change the prediction accuracy. In the end, the machine learning model and analysis tool work together to provide not only a stronger analysis of the LULC-LST relationships but a more human-interpretable result (via SHAP values), which overcomes the largest challenge in current machine learning techniques of result interpretation difficulties.

2. Materials and Methods

The methods of this study follow a similar hypothesis of previous research [23,35,39] believing that vegetation coverage, water bodies/moisture, impervious surface, and park characteristics have an impact on UHI and that the cooling effect of parks is also dependent on the land surface types outside the park spaces [37,40]. Many researchers employed the thermal conduct theory, a physical science methodology, to investigate the heat balance between the park and its surrounding area [41–43]. This method of study can assist in providing a methodical approach to understanding the cooling effect of parks. However, departing from previous studies, we utilized machine learning techniques to provide a more robust study on the relationship between our chosen factors and the park cooling intensity and land surface temperatures to see which has the higher impact. More specifically, a boosting model was used in the analysis as it could handle null values as well as large datasets. All machine learning algorithms, including boosting models, have the potential to overfit, however, standard multivariate linear regression is guaranteed to overfit due to Stein's phenomena or as some refer to as Stein's Paradox. This phenomena states that estimating the maximum likelihood of an average is suboptimal. In Stein's works, he proved that by shrinking the maximum likelihood estimator, it is possible to get better performing estimators. In machine learning modeling, this is not necessarily a new concept but is typically referred to as regularization [44].

2.1. Study Area Selection

Dallas (32.7767° N, 96.7970° W) is the largest city in the North Central Texas region. This metropolitan area is situated on a coastal plain, about 400 km north of the Gulf of Mexico. The city and surrounding region has been the center of major economic and demographic structural changes over the past few decades, most notable between 2000 and 2010 seeing a nearly 23% increase in population density with the region projected to be a "megacity" with 15 million inhabitants by the year 2050 [20]. This has led to rapid industrialization and urbanization of the city and surrounding areas, mainly in the form of urban sprawl.

The growth of the city has brought about a greater density of impervious surfaces, mainly in physical infrastructure, parking lots, buildings, and commercial/industrial areas. With this growth, there has also been a noticeable increased air pollution and energy consumption, and a consistent rise in the UHI. Following with the theories stated in the introduction, the UHI of Dallas has been influenced by the local land surface usage and land coverage, regional climate variations, large-scale climate variability, and long-term climate change due to meteorological variables, like cloud cover, wind speed, and soil moisture [20].

As much of the previous research has focused on cities that have grown denser and taller, Dallas makes a great sample city to see how the same theories apply to cities that have urbanized but have also grown outward.

2.2. Data Acquisition and Analysis

2.2.1. Data Sources

In this research, land use/land cover data from North Central Texas Council of Governments (NCTCG) was obtained for Dallas, Texas. Additionally, Landsat 8 satellite images from the United States Geological Survey (USGS) were obtained at a resolution of 30 m. The satellite images were an aggregation of 10 images taken over the summer of 2015 in the city to match the county level land use/land cover dataset dates provided by NCTCG.

We chose to focus on summer months as other researchers have found that the UHI intensity and ranges are significantly higher in the summer months when compared to other seasons [11,45]. This higher range and intensity are helpful when performing impact analyses due to the ability for the model to easier assess the impact of the independent variables.

2.2.2. Land Use and Land Cover Dataset

Our dataset for land use/land cover was provided by NCTCG via their open data portal [46]. The NCTCG releases an updated dataset every 5 years for land use/land cover data. At the time of this research, the most recent publicly available dataset, and our chosen dataset, was for the year 2015. Thus, our park's dataset was constrained to match this temporal frame having parks only built before 2015; in addition, there was a similar temporal restriction on our satellite images collected, which was used for calculating LST.

In total there were 33 separate LULC categories that were aggregated, with guidance from the NCTCG dataset documentation, into 11 separate groups, Appendix A, being: Residential, Commercial/office, Industrial, Misc. built-up, Transit, Mixed-use, Parks/recreation, Misc. open (green), Misc. open (non-green), Flood control, Parking, and Water.

More specifically—the Misc. built-up is an aggregation of the utility, airport, runway, large stadium, railroad, and communication categories; Misc. open (green) is an aggregation of vacant land, residential acreage, ranch land, timberland, and improved acreage categories; and lastly, Misc. open (non-green) is an aggregation of landfill, construction site, and cemetery categories.

The park location data and polygons were constructed from the parks and recreation category of the NCTCG dataset. The size of the parks within the city varied largely with the largest being over 9 million square meters and the smallest being around 30 square meters. To remove any outliers in the dataset related to park size, we utilized a z-score method. As the z-score is valuable to represent how many standard deviations away the given value is from the mean value, using a z-score threshold of ± 3 allowed us to remove any data points where the park's size was larger or smaller than 99.7% of the data. After performing that transformation, we were left with a total of 1201 parks ($n = 1201$). Overall, the parks used in our dataset had an average of 380,660 square meters. As seen in Figure 1, the distribution of the parks follows more of a natural path going along rivers and around lakes; however, there is still a decent scattering of parks in and among the suburbs of the city.

Figure 1. Dallas, Texas with parks overlaid and example park displaying 50-, 100-, 150-, 200-, 250-, and 300-m buffer zones.

The parks' perimeter acted as the center point for our buffers, created in ArcGIS, at 50 m, 100 m, 150 m, 200 m, 250 m, and 300 m (Figure 1). The buffer interval (50 m) is a recommended choice as it is almost double the size of our satellite image resolution (30 m) used to calculate our LST data, as mentioned in Section 2.2.3 [47]. We chose a 300 m limit to match with previous research, finding that the extent of the parks' cooling effect does not extend much past that. As well, 300 m has been suggested as the limit to define a "recreational neighborhood" based on how far people are willing to walk. This means that outside of that 300 m buffer distance, the park is no longer considered to be part of the recreational zone [47–50].

Lastly, within each of these buffer zones, we calculated the area in square meters occupied by each of the 11 separate LULC classifications that were constructed from the NCTCG dataset.

2.2.3. Retrieval of LST and Average LST Calculations

To retrieve the Land Surface Temperatures (LST) in the city, we utilized the methodology of the Remote Sensing Lab under the Institute of Applied and Computational Mathematics (IACM) [51]. The lab provides open-source code, which gives us the ability to easily obtain the LST data for our study area. The researchers were able to accurately predict LST with a RMSE of 1.52 °C.

Following the methodology proposed by the IACM, the data has been precision ortho-corrected; corrected for brightness temperature and cloud mask, Landsat surface reflectance, and emissivity; and has undergone atmospheric correction to provide the highest level of accuracy and standardization.

As mentioned, the satellite images we chose for our analysis were captured from the Landsat 8 satellite at a 30 m resolution. With the lab's methodology, we were able to create 10 separate LST datasets for Dallas over the summer 2015 season, which corresponds with our LULC data date range. These different datasets were averaged together, utilizing ArcMap's raster calculator, to help with managing limitations of lack of meteorological data as the average of the 10 datasets is a more rounded depiction of the LST than a single dataset from a single satellite image.

After the creation of our LST dataset, the data was further processed in ArcGIS to calculate the average LST within the park borders as well as within buffers around the park at each 50, 100, 150, 200, 250, and 300 m radii. These calculations were then used in the computation of PCI (Section 2.2.4).

2.2.4. Determination of the Park Cooling Intensity (PCI)

For Park Cooling Intensity (PCI) analysis studies, there are two main approaches to data collection. The first being in person field observation to collect data from the study area manually; and the second being data from monitoring stations or through thermal infrared (TIR) remote sensing [52,53]. For this study we chose the latter, utilizing remote sensing. There is a growing availability of high-quality satellite images, taken at regular intervals, over large areas of land. Additionally, there have been improvements in LST retrieval algorithms allowing for more accurate characterizations of the urban thermal environment [54–56]. As previously mentioned, for our specific study, our LST retrieval algorithm follows the methodology of the Remote Sensing Lab under the Institute of Applied and Computational Mathematics (IACM) [51] due to its robustness and flexibility.

When calculated, PCI (1) shows the temperature difference between the inside of the park and the urban area immediately bordering it [57,58]. In this study, the PCI (measured in °C) used mean LST in the calculation as opposed to air temperatures.

$$PCI = Tu - Tp \tag{1}$$

where Tu refers to the mean LST of the urban area at a buffer zone outside of the park; Tp indicates the mean LST inside the park's borders.

The buffer zone includes the area around the park, which contains different LULC types: industrial/commercial areas, buildings, roads, etc. (impervious surfaces), and trees and other green spaces.

2.2.5. Dataset Construction

After the collection of the land use/land cover data and Landsat 8 satellite images, it was possible to construct our dataset to be analyzed. As mentioned in Section 2.2.2, our park data was created from the LULC data collected from the NCTCG using the parks and recreation LULC category to place where the parks in the city are and get their geometry as a shapefile. With the park geometry and location data, it was then possible to calculate the 50, 100, 150, 200, 250, and 300 m buffers using ArcGIS.

Land surface temperatures (LST) were calculated as outlined in Section 2.2.3 utilizing a third-party methodology and open-source Python code. With the buffers and LST calculated, it was then possible to calculate the park cooling intensity (PCI), outlined in Section 2.2.4.

After calculating the park geometry, buffers, LST, and PCI we were then able to complete our dataset by getting spatial statistics of the mean PCI in each buffer around the park, the mean LST, and the proportion of space (in square meters) taken by each LULC category making up each buffer zone. Our data construction framework can be visualized best with Figure 2.

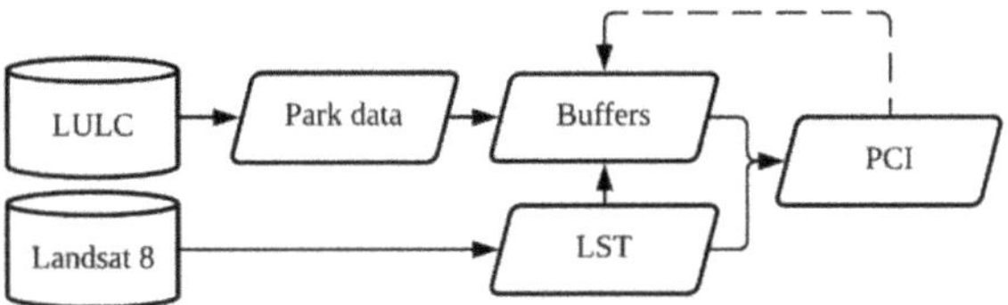

Figure 2. Framework for dataset construction.

2.3. Analysis Methods

As previously mentioned, our methodology utilizes machine learning techniques. For our model, Extreme Gradient Boosted (XGBoost) machine learning models were chosen to produce SHAP values that can be used to accurately analyze both the effects of the various LULC classifications on the PCI and LST at the diverse buffer distances.

Our model's framework, visualized in Figure 3, is as follows, with a_i and r_i making up the regularization parameters, i being the respected regression tree, and h_i being a function trained to predict residuals (r_i) using X. XGBoost computes a_i by using the computed residuals and the following Equation (2).

$$\min a = \sum_{i=1}^{m} L(Y_i, F_{i-1}(X_i) + ah_i(X_i, r_{i-1})) \text{ with the loss function being} : L(Y, F(X)) \quad (2)$$

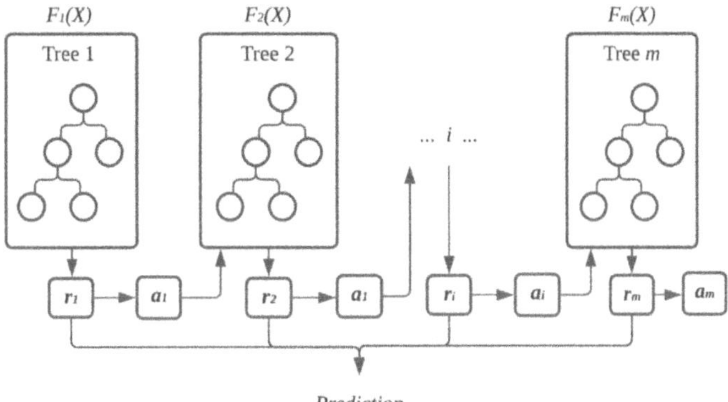

Figure 3. Framework for our XGBoost model.

As we are using XGBoost for a regression analysis, the weak learners are regression trees. To get the best estimate, the model makes multiple smaller, and weaker, estimates (regression trees) and then averages their residuals to get the final estimation. The training proceeds iteratively, adding new trees that predict the residuals or errors of prior trees that are then combined with previous trees to make the final prediction. It is called gradient boosting because it uses a gradient descent algorithm to minimize the loss when adding new models.

2.3.1. XGBoost

With our dataset, constructed in Section 2.2.5, we began constructing our XGBoost model using Python and the XGBoost package (version 1.5.0).

The focus of this analysis was to determine which LULC classification has the largest impact on the land surface temperature and PCI, to give us a better understanding of which of the classifications have the highest impact on changing temperatures and a better understanding on which classifications have the highest impact on parks' ability to affect temperatures.

XGBoost is a highly flexible, quick, and versatile algorithm that can work through regression, classification, and other problems. It is the evolution from decision-tree based models and uses a gradient boosting framework. For a more detailed breakdown of the model, please refer to Chen's study [59] on the creation of the algorithm.

We selected this model due to its high performance and accuracy in preliminary tests as well as for its scalability. With each buffer having varied levels of each of the LULC classifications, we needed a model that could be comparable across each of our buffer zones and a model that could be accurate without the need to construct a new model for each buffer zone. Additionally, XGBoost can handle null values well, on issues that we faced in our dataset (for example, very few 50 m buffers have water and there exists an overall lack of mixed-use LULC classification type among all the buffers).

For the performance analysis, the Root Mean Square Error was employed due to the noisiness of the data as an attempt to add extra weight to large errors.

For the importance modeling analysis, the main scope of the study, we borrowed a game theoretic approach known as SHapley Additive exPlanations (SHAP). SHAP can give a detailed and clear picture of which factor has the largest impact on the model by removing each factor one at a time and trying a combination of factors to see how the combinations change the prediction accuracy.

Our model, as shown in the framework above, followed a 70/30 split on the dataset created in Section 2.2.5. To calculate our model's hyper parameters, we did not utilize the popular grid search method and instead opted for the speedier Bayesian Optimization. For models and datasets that can be computationally heavy, the grid search method can be equally heavy and time-consuming. Bayesian Optimization, on the other hand, can find the optimal parameters by learning from previous optimizations. In the end, this method has greater success but is also less computationally expensive and quicker. Using this method, we calculated the parameters shown in Table 1. Lastly, our model was then trained and tested using Python and XGBoost.

Table 1. Parameters calculated using Bayesian Optimization used in our model.

Parameter	Value
colsample_bytree	0.5
eval_metric	rmse
min_child_weight	30
subsample	0.7
eta	0.075
objective	reg:log
max_depth	9
booster	gbtree

2.3.2. SHAP Values

SHAP values, founded by Lundberg and Lee [60], are best suited for when you are processing a more complex model, such as gradient boosting or neural networks, and are needed to better understand what decisions the model is making. By using predictive models alone, we can answer questions related to "how much" change takes place, but with the addition of SHAP values we are able to go a step further and begin to answer "why" the changes are happening.

To more easily explain our XGBoost model, we provide summary plots for the SHAP values produced. Each plot contains the LULC category (left hand y-axis), a colormap of the feature value of the individual datapoint (right hand y-axis), and the SHAP value for the individual point (x-axis). These summary plots are similar to the partial dependence plots; however, they have an advantage with their ability to display the extent to which the interaction terms matter.

It is important to note that while these plots explain our XGBoost models, it does not necessarily represent a causal relationship. Since our XGBoost model is trained on observational data, the model is not necessarily attempting to observe causation. With that, the results are not stating causal relationships, and stating an increase in one variable will lead to a change in another; however, the results do present the factors with the highest impact on the models' predictions.

A main consideration for choosing SHAP as our main tool for analysis is that it is possible to see the overall impact of each datapoint as well as the impact of null data points, which is a similar consideration when choosing to use XGBoost. Traditionally, when there is a point missing, it is replaced with a 0 or an average of similar values; however, in some cases, this can tamper with the data. In our case, replacing the values with a 0 would have skewed our data significantly due to the large number of missing values for some classifications in some buffers; in addition, providing an average number would have skewed our results in the opposite direction and would have taken away the game theory aspect of SHAP.

3. Results

To better understand which land use/land cover classifications had the greatest impact on LST and PCI, we used SHAP values. A SHAP value represents how much the model prediction changes when we observe a certain feature, in the case of this study, the area size of each LULC classification. In the summary plots throughout this section the SHAP values for a single feature are plotted on a row, where the x-axis is the SHAP value. With this method, it is possible to see which features drive the model's prediction the most, as well as which only affect the prediction a little. As well, in each plot, the individual dots are colored to represent the value of the feature from high to low.

3.1. Relationship between LST and LULC Classifications

The below subsections present the results of the analyzed relationship between park LST and the LULC classifications at each buffer level. While the buffers were at distances of 50, 100, 150, 200, 250, and 300 m, for simplicity the following section will present results at only three separate buffer zones, 100, 200, and 300 m. However, the section will be ended with an overall analysis of all six buffer zones.

To better explain the models, the SHAP values were normalized and summed to 100 for each plot to represent 100% of the model. However, the SHAP values on the summary plots themselves are not normalized and are the true SHAP value for each of the models.

Figure 4 below shows each individual datapoint value compared to the overall SHAP score for the specific point and its LULC classification when modeling for impact on LST. The right y-axis in the figure shows the feature value, which indicates how high or low the specific buffer's average LULC area is (with grey features being null values). On the left y-axis, we can see the individual LULC classifications. And, lastly, on the x-axis we can see the SHAP value with negative values indicating that the LST is decreased and positive, indicating that LST is increased.

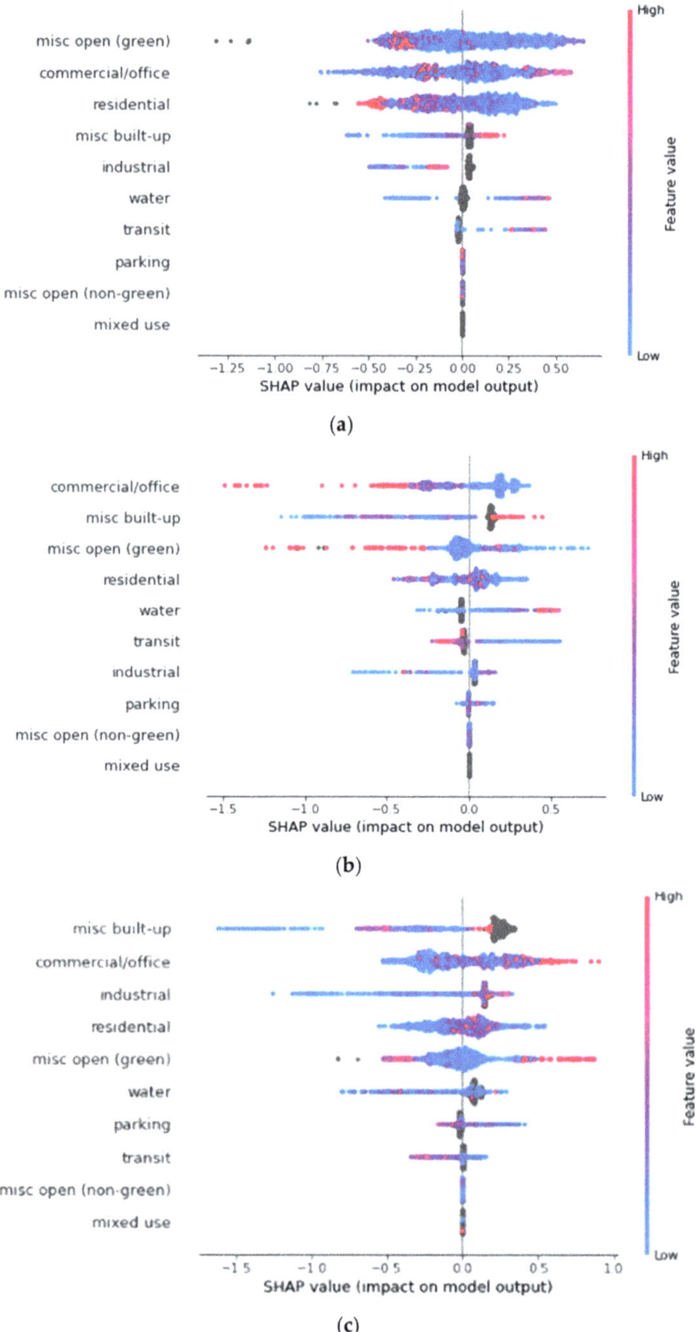

Figure 4. The SHAP values from the XGBoost model at the different buffer zones around the parks for the impact analysis of each Land Use/Land cover classification group on Land Surface Temperature. (**a**) 100 m buffer, (**b**) 200 m buffer, (**c**) 300 m buffer.

3.1.1. 100-m Buffer

At the 100-m buffer zone, visualized in Figure 4a, our model returned an RMSE of 1.56. With the minimum LST being 31.49 °C and the maximum being 45.89 °C, we have a range of 14.40 °C, with that consideration the RMSE of 1.56 is decent and shows that the model was able to predict the average temperature of any individual 100-m buffer zone within 1.38 °C. With this current model, "misc. open (green)", commercial/office, and residential explain 73% of the impact on LST.

In Figure 4a, "misc. open (green)" has the largest impact on the model overall, explaining roughly 26.32% of the model. Based on these results, there is a negative relationship between this LULC category and LST. The results show that an increase in area coverage of the "misc. open (green)" classification within the 100-m buffer distance of the part has a greater chance to negatively impact LST. Likewise, it shows if the area size of "misc. open (green)" within the 100 m buffer is lower, then there is a greater chance that that area has a higher LST. These results are in line with others that have found green spacing to have a cooling effect on the surrounding areas [35,36,38,61] as well as case studies that have found the greening of cities as a significant source of UHI mitigation [5,25,27,33].

By using SHAP values, we can see which factors have the highest chance to increase (the right of the 0.00 mark) or decrease (the left of the 0.00 mark) our dependent variable. As well, the intensity, or level of impact, is represented to show how high or low the feature value is. A high feature value indicates a large presence of the specified feature and vice versa with a low feature value indicating a smaller presence.

The next most impactful feature at the 100-m buffer was the "commercial/office" category, explaining 24.54% of the model, having a more mixed result. With "commercial/office" zones, we can see that there is a mixed effect of high values potentially raising or lowering the LST. However, higher values have a higher impact factor when it comes to raising the LST indicated by the higher SHAP values for 'high' feature value data points. When analyzing SHAP values, it is also important to consider the low values and while the case of the commercial/office zones still looks quite mixed, the slight tail of low values going negative does hint that a lower total area of commercial/office zones has a higher impact on decreasing LST. Considering both sides of this, we conclude that at the 100 m buffer distance, commercial/office zone area has a positive impact on LST, in other words, the greater the area of this zone, the greater the LST.

The third most impactful feature was "residential" zones explaining 22.86% of the model. For residential zones, after analyzing the plot, we conclude that, at the 100-m buffer distance, this zone has a positive impact on lowering LST as the higher feature values are to the left of the plot and lower feature values are to the right. We suspect that this is due to the large amounts of suburban parks in the city of Dallas. As suburbs are typically "greener" with yards, tree lined streets, and more green areas overall, they are more likely to act a semi-green/semi-built-up area, have less of an impact on increasing LST, and in this study appear to be collaborating with the parks in lowering LST.

As seen by previous studies, "built-up" areas are more likely to contribute to increased LST [39,42,45]; this can be a partial explainer of the "fuzziness" in the interpretation of the "commercial/office" zones as well as the decreased impact of residential areas as they are generally a mix of both built-up and green spaces that have been shown to have competing effects on LST.

Following these features comes misc. built-up explaining 7.66% of the model with a positive effect on LST; industrial explaining 7.43% and having an overall negative impact on LST but a trend of more area coverage of this type increasing LST; water explaining 6.17% of the model with a positive effect on LST, which we did not expect; transit explaining 5.02% of the model with a positive impact; and, lastly, parking, misc. open (non-green), and mixed-use explaining <0% of the model without any measurable effect.

3.1.2. 200-m Buffer

At the 200-m buffer zone, visualized in Figure 4b, the same trend from the 100-m buffer continues overall. The RMSE for this model was 1.79 with the min temperature being 31.47 °C and the max being 45.98 °C, making a range of 14.5 °C.

The largest change in increasing the buffer size is that the number of data points also increase, allowing new trends to emerge. Here, the biggest change in impact observed is the somewhat large increase in the impact of "misc. built-up" areas on LST. At the 200-m buffer zone, the misc. built-up feature area size accounts for 21.23% of the model, an increase from 7.66% at the 100 m buffer. This LULC category has a positive trend with LST, with a large tail of low values. The absence of this feature can be associated with areas of lower LST and in areas that this category is present there are increases in LST. This is in line with previous research that has found that the increase of impervious and built-up areas as well as the decrease in trees and vegetation as contributing factors increase LST [29,33,39,45,56].

Like the 100 m buffer analysis, the commercial/office classification is one of the more impactful features, explaining 25.14% of the model with a quite clear negative impact on LST, differing from the 100 m buffer zone result. The reason for this shift is not explainable with SHAP values alone and would require further and more in-depth research into the specific makeup of the commercial/office zones in the City of Dallas, something that is outside of the scope of this study.

The third most impactful feature is misc. open (green), explaining 15.28% of the model. The trend here maintains and strengthens from the first model at the 100 m buffer. This is followed by residential (13.20% of the model) with an ambiguous impact; water (9.43% of the model) with a positive impact; transit (7.52% of the model) with a stronger negative impact on LST, differing from the 100 m buffer analysis; industrial (7.03% of the model) with a negative impact on LST; parking (1.17% of the model) with an ambiguous trend; and then misc. open (non-green) and mixed-use having no trends and explaining 0% of the model.

3.1.3. 300-m Buffer

Our last buffer analysis is at the 300-m zone, visualized in Figure 4c, the furthest out and the one with the most data points. The RMSE for this model was 1.92 with the min temperature being 31.58 °C and the max being 46.11 °C, with a range of 14.53 c.

This buffer zone analysis saw the misc. built-up LULC classification move to the top as the most impactful feature, accounting for 25.81% of the model. For this classification, there is a strong trend showing lower area coverage of misc. built-up areas having a higher negative impact on the model's prediction. This is followed by commercial/office, having a positive trend on LST, accounting for 19.27% of the model. The third most impactful feature in the 300 m buffer zone was industrial, accounting for 16.16% of the model and having a strong positive impact on LST with lower feature values having a much higher impact on the dependent variable than higher feature values.

The next most impactful feature in this buffer was residential, making up 12.40% of the model. Again, the residential values seem to show less of a trend. One plausible explanation for this is the large and varied types of residences in the city, with this one classification comprising suburban homes, apartment buildings, and even mobile homes.

After residential LULC classifications were misc. open (green) spaces closely following with 11.57% of the model's explanation. This feature appears to show that high feature values can either increase or decrease the LST of an area, and lower values seem to have no effect. As this feature is a "misc." feature and acts as an aggregator of smaller classes, this type of result requires more research to better determine which specific features of misc. open (green) are contributing to either the decrease or increase of LST.

Water (with 9.30% of the model) continues to show a positive impact on LST with lower values having a higher impact on lowering LST. This is followed by transit with 1.99% of the models' impact, and misc. open (non-green) and mixed-use with no trend and 0% of the model's explanation or not impactful on this specific model.

3.1.4. LST SHAP Comparison

In this section, we combine each of the LULC classes and the average SHAP values across all the points (1201 data points for each model) in each model to both compare the trends among all six buffer zones and then aggregate the SHAP values together as an average to get an idea of the overall impact of each class. To compare the SHAP values between groups, the values were normalized so the values for each of the models sum to 100 as SHAP is not naturally comparable across models.

Figure 5 shows the trends for each of the different buffer zones across each of the LULC categorise. As displayed, each buffer follows roughly the same pattern with "misc. open (green)" and "misc. built-up" being the two with the most similar trends across buffers. This does not come as a surprise as both built-up and green areas have been heavily studied to show their impacts on LST, as cited throughout the paper.

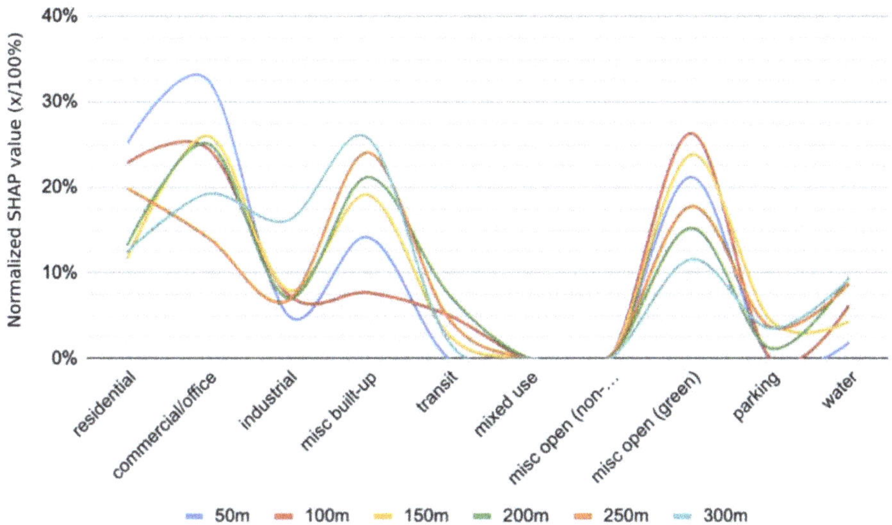

Figure 5. SHAP values by buffer for LST analysis. Averaged normalized SHAP values (x/100%) for each LULC classification group and buffer zone distance to show the overall trend in the impact of each group on LST by buffer.

Additionally, the residential and commercial/office are generally a hybrid of built-up and green spaces making their impacts on LST a bit more difficult to measure; however, they do still present a significant impact [3,8,30,56]. This difficulty in measurement explains their less resemblant trends as there are a multitude of interactions taking place within these categories as well as between them and others.

The "misc. open (non-green)", "mixed-use", and "parking" were the least impactful variables across all buffer areas. Notably, for mixed-use, this lack of impact is likely due to the lack of actual mixed-use zoning in the City of Dallas. We also suspect that parking was not as prominent as expected due to it being naturally grouped with other areas such as "commercial/office" and rarely standing on its own.

To give a better idea of the aggregate trends, the SHAP values were averaged to produce Figure 6. In this figure, "commercial/office" is shown as the most impactful LULC category, accounting for 23.61% of the model's power. From the individual buffer zone breakdowns showing this category having a positive impact LST, we can assume that, overall, that a higher area ratio of "commercial/office" space has the potential to lead to higher LST. However, as mentioned before, this is a difficult assessment to make due to the makeup of these zones, especially in a city like Dallas that has experienced significant urban sprawl. The category with the next highest impact "misc. open (green)" having 19.34% of the prediction power and, again based on the buffer zone breakdowns, has a stronger trend towards decreasing LST. This category is closely followed by "misc. built-up" with 18.69% of the model's prediction power and, based on the breakdowns, also shows an overall trend of increasing LST. These results of green spaces and built-up areas having large, but opposite, impacts on LST follow results found by other researchers who have found that green areas have a cooling affect while impervious/built-up areas contribute to urban warming [23,27,37,39,40,47].

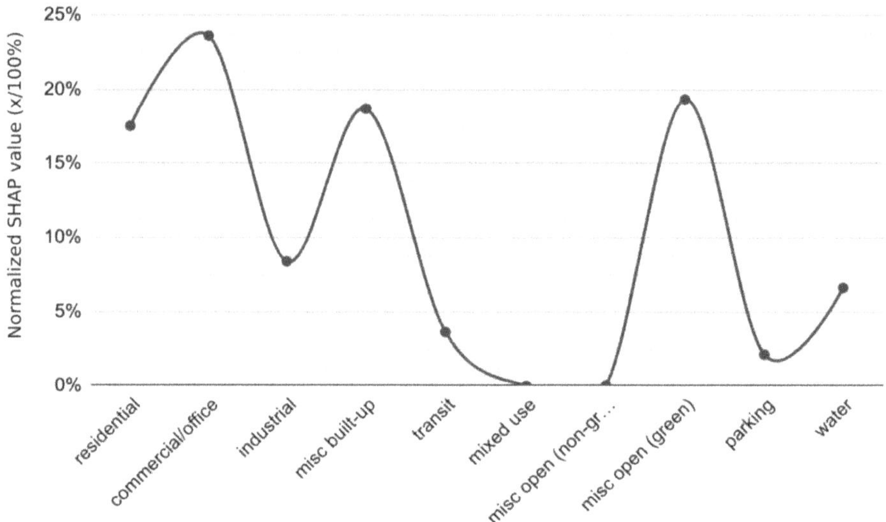

Figure 6. Aggregate SHAP values for LST analysis. Further aggregation of averaged normalized SHAP values (x/100%) for each LULC classification group show the overall trend in the impact of each group on LST.

The remainder of the variables accounted for less than 10% of the model's prediction power with "industrial" having 8.38% (increasing LST), "water" having 6.62% (no trend), "transit" with 3.64% (no trend), "parking" with 2.13% (no trend), and lastly "mixed-use" with 0% (no trend).

3.2. Relationship between PCI and LULC Classifications

PCI is typically defined as the temperature difference between inside the park area and the urban areas within a set buffer. It gives insight into how much colder the park area is compared to the buffer regions and can give an idea of how a park cools its surrounding areas. Like Section 3.1, we performed the same analysis while replacing LST with PCI to get a better idea of how the various park characteristics affect the temperatures of the areas around them and not just inside the park itself. If PCI is 0.00, then that means the buffer zone is the same temperature as the park; if the value is positive, then it means the buffer zone temperature is higher when compared to the park, and likewise, when the value is negative it indicates that the buffer zone temperature is lower than the park.

Similar to Figure 4, Figure 7 below shows each individual datapoint value compared to the overall SHAP score for the specific point and its LULC classification when modeled for impact on PCI. The right *y*-axis in the figure shows the feature value, which indicates how high or low the specific buffer's average LULC area is (with grey features being null values). On the left *y*-axis, we can see the individual LULC classifications. Lastly, on the *x*-axis we can see the SHAP value with negative values indicating that the PCI is decreased, and positive indicating that LST is increased.

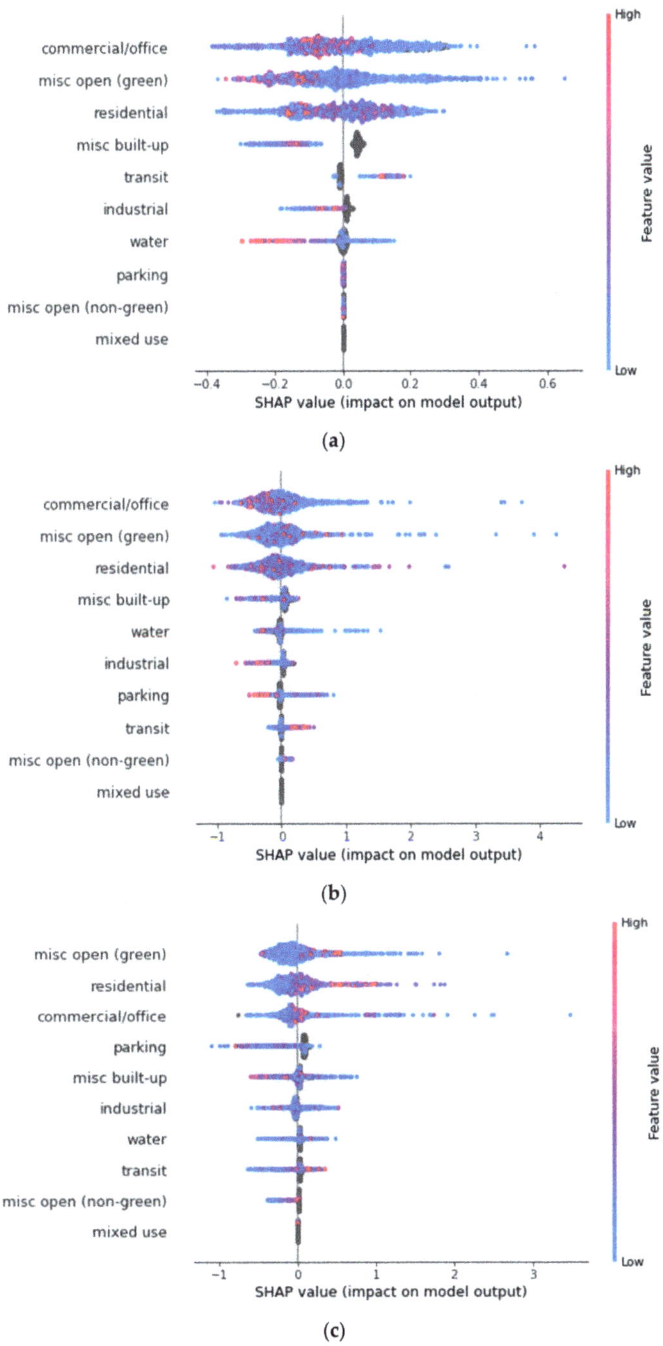

Figure 7. The SHAP values from the XGBoost model at the different buffer zones around the parks for the impact analysis of each Land Use/Land cover classification group on Park Cooling Intensity. (**a**) 100 m buffer, (**b**) 200 m buffer, (**c**) 300 m buffer.

3.2.1. 100-m Buffer

At the 100 m buffer zone analysis, visualized in Figure 7a, our model had an RMSE of 1.37. For the 100 m buffer zone, the minimum PCI was −1.128, indicating that, at least one park buffer had a temperature decrease of 1.13 °C in comparison to the park; as well, there is a max PCI of 18.99, indicating an 18.99 °C increase in temperature of at least one of the buffer zones in comparison to the park. The range between the min and max PCI is 20.13, meaning the RMSE for this model is quite good. As would be expected, the LULC classification with the highest impact on the model is similar to those in the LST analysis. However, this time, we are not considering how the feature impacts LST directly but how it impacts the PCI. In other words, we are interested in seeing which LULC class has the biggest impact on increasing and decreasing the cooling effects of the parks.

In the 100 m buffer zone, the "commercial/office", "misc. open (green)", and "residential" classes had the largest impacts making up over 75% of the model's prediction power with 26.36%, 26.13%, and 22.19% shares of the model respectively. Similar to the LST analysis, commercial/office shows a slight impact on decreasing the PCI at the 100 m buffer range. As well, misc. open (green) has a heavy impact on decreasing PCI. This means that both factors, at the 100 m buffer zone, help with the park's cooling effects.

For residential zones, there is less of an obvious trend for when the area ration of this class increases; however, the plot does indicate that as the area ratio is decreased there is a slight negative impact on PCI.

In the same model, "misc. built-up", making up 13.97% of the model, is also shown to have a negative impact on PCI with the plot indicating lower values of this class leading to a higher impact on lowering the PCI. This trend is also the same with "transit" (with 3.84% of the model) and "industrial" (3.79% of the model), with all showing a decrease in PCI with a decrease in their feature value and an increase in PCI with an increase in their feature value, or some of both relationships together.

The LULC class here that was very different from the LST analysis was "water". While it shared only 3.73% of the model's prediction power, it also showed a very negative impact on PCI. For the model run, the water class was shown to decrease the PCI as its area ratio increased, and the PCI increased as its area ratio increased. The remaining classes had no impact on the model's prediction power.

3.2.2. 200-m Buffer

This model, visualized in Figure 7b, gave an RMSE of 1.62, with a min PCI of −1.75 and a max PCI of 22.24, having a range of 23.98. This result was interesting as the results seem to be driven by the lack of a LULC classification rather than the presence of it, and the results are slightly different than the others. However, this model had very similar SHAP values as the 100 m buffer analysis.

For this buffer, the "commercial/office" LULC classification had the highest impact on the model, holding over a quarter of the model's prediction power at 25.44%. While the trend is slight, there does appear to be a negative impact on PCI with the commercial/office area ratio. This result is the same with the "misc. open (green)", which has 23.14% of the model prediction.

Unlike the 100 m buffer analysis, "residential" shows a slight positive impact on PCI, meaning that the residential zones can potentially increase the PCI. However, from the plot, this relationship is not obvious. As previously mentioned, the residential class is a bit broad as Dallas has a wide range of housing types (from suburbs to multi-family apartments) that are all grouped into one class.

"Misc. built-up" (8.65% of the model) has a slightly mixed result that indicates the possibility of this class to negatively impact PCI when the feature values are too high or low. This can be due to a limited data size, or due to some other factors involved in this class.

The "water" class (6.13% of the model) continues to show a negative impact on PCI, however, with a higher impact on PCI increasing as the area ratio of water is lower.

In this buffer analysis, the "industrial" class (5.75% of the model) gave a bit of a different result with a plot indicating a negative impact on PCI. This is the same with the "parking" class (5.69% of the model). However, the "transit" class (2.33% of the model) shows a heavy trend in increasing the PCI at this buffer level.

The "misc. open (non-green)" class, while showing a negative impact, only accounted for 0.34% of the model and "mixed-use" accounted for 0%.

3.2.3. 300-m Buffer

At the 300-m buffer zone, visualized in Figure 7a, our model has an RMSE off 1.59 with the min PCI being −1.94 and the max being 23.15, a range of 25.09, indicating a good RMSE. This model, having the most data points, also produced some of the clearest results for the factors that impact PCI the most.

For this model, the most impactful LULC class was "misc. open (green)", accounting for 21.19% of the model. This feature shows a negative impact on PCI due to the higher PCI values when the feature value is lower. The "residential" class had the next highest impact, sharing 19.73% of the model. The residential feature showed a clear positive impact on the PCI, meaning that higher area ratios of residential zones led to a higher PCI in the 300 m buffer.

After residential was the "commercial/office" class, with 16.26% of the model, also showing a positive impact on PCI. The next feature, "parking" with 13.63% of the model, followed a similar trend and showed an overall positive impact on PCI. However, with parking, there are a couple of rows of data that show a lower PCI with higher parking area ratios. To fully determine what would cause this would require a more robust dissection of the parking LULC class, which is beyond the scope of this study.

One odd result here was that the "misc. built-up" class, holding 8.42% of the model's prediction power, showed a very slight negative impact on PCI. This goes against the LST analysis, which showed the opposite. The "industrial" class made up 7.29% of the model, with no real trend; "water" made up 5.89% of the model, again with no real trend; then finally, "transit" and "misc. open (non-green)" both showed a positive impact on PCI, making up 5.81% and 1.79% of the model respectively.

3.2.4. PCI SHAP Comparison

Similar to Section 3.1.4, the various SHAP values and LULC classes were compared to try to determine any overarching trend. Unlike the similar process done for the LST analysis, the PCI analysis does not have as many similar patterns. As seen in Figure 8, the major similarity between the six buffer zones is the "misc. open (green)" LULC category and to a lesser extent the "commercial/office" and "misc. built-up" classes. This shows that despite the distance from the park, the presence of additional green spaces had a significant impact on the PCI as did residential and commercial/office areas. Interestingly, the built-up and industrial categories had lower impacts the closer to the park they were, and increased in impact until about 150 m away and then the increase became less linear. The best explanation for this is first a potential lack of data, but secondly it is supported by the ideas belonging to thermal conduct theory research and investigations into the heat balance between the park and its surrounding area especially in relation to impervious and built-up areas where it has been found that heat can transfer from these surfaces to parks more easily at the edges of the parks where they come into direct contact. [41–43].

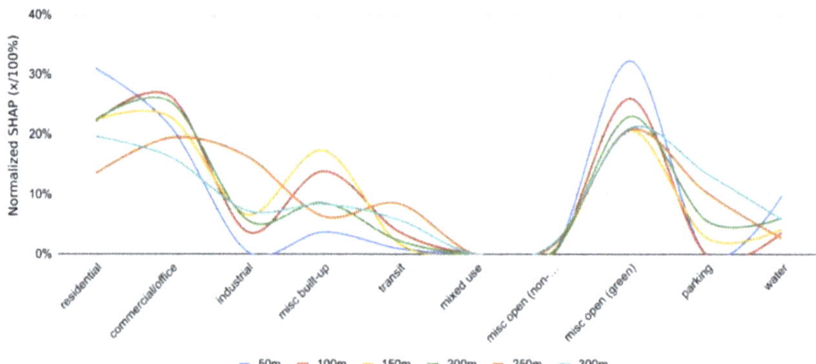

Figure 8. SHAP values by buffer for PCI analysis. Averaged normalized SHAP values (x/100%) for each LULC classification group and buffer zone distance to show the overall trend in the impact of each group on PCI by buffer.

Again, as done in Section 3.1.4, we have aggregated all the SHAP values together to see the average trend among the buffers (Figure 9). When comparing Figures 6 and 9, the overall trends are nearly the same when modeling LULC categories on LST and PCI with Figure 9 having softer curves. However, with the PCI SHAP averages, "misc. open (green)" spaces show the highest impact on PCI with a 24.11% average and an overall negative trend on PCI, meaning that higher area ratios of "misc. open (green)" contribute to lower PCIs and aid in the park's cooling, which is supported extensively in previous literature on the topic [16,27,37,38,40,47,61]. Commercial/office and residential are nearly tied with 21.98% and 21.93%, respectively, and both have a slight negative effect on the PCI. As mentioned, and supported by previous research, commercial/office and residential zones are generally a hybrid of built-up and green spaces, making their impacts on temperatures and PCI a bit more difficult to measure; however, they do still present a significant impact [3,8,30,56]. These values are followed by misc. built-up with 9.77% (a very slight positive effect, or possibly no effect), industrial with 6.74% (a positive effect), parking with 5.44% (a positive effect), water with 5.44% (a negative effect), transit with 3.90% (a positive effect), misc. open (non-green) with 0.68% (no effect), and mixed-use with 0.00% (no effect).

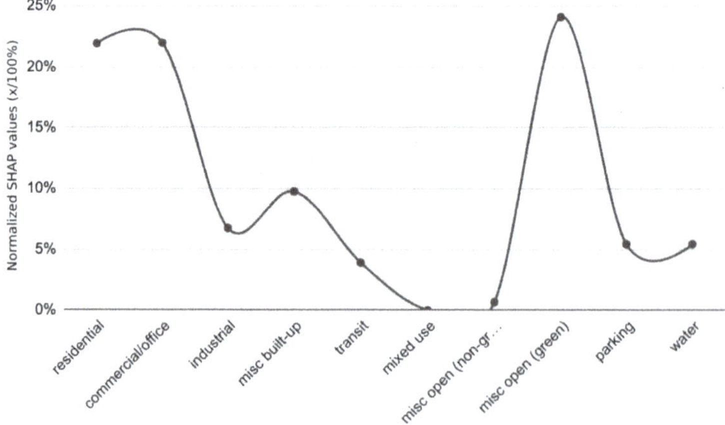

Figure 9. Aggregate SHAP values for PCI analysis. Further aggregation of averaged normalized SHAP values (x/100%) for each LULC classification group to show the overall trend in the impact of each group on PCI.

3.3. PCI at Different Buffer Distances

While our research aims to study the impact of each LULC grouping on PCI, it is also important to analyze the extent of the cooling effect. The PCI values were aggregated at each of the buffer distances (50, 100, 150, 200, 250, 300 m) to determine any relationship between the distance and PCI. As seen in Figure 10, there is an increasing trend in PCI across the buffers.

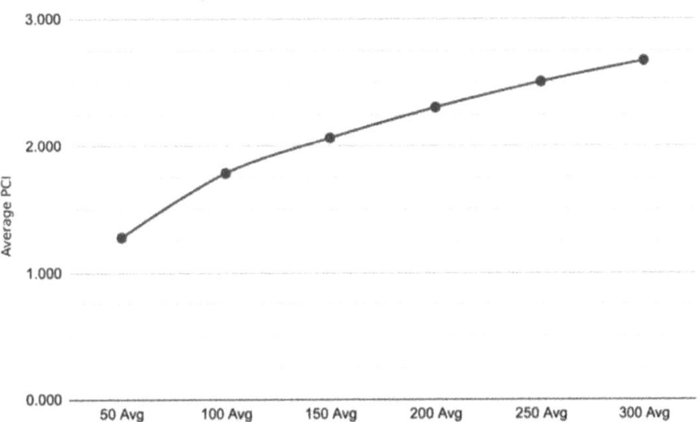

Figure 10. Average PCI at various buffer zones around parks. The average PCI at each buffer zone shows the change in PCI as the distance from the park changes.

It can be determined that the parks do have a cooling effect on the surrounding areas, indicated by the PCI increasing as the distance from the parks increases. The largest increase in PCI is from the 50 m to 100-m buffers and then the trend becomes more linear. Previous research found similar results showing that parks had a positive relationship with PCI intensity up to 300 m buffer region [47]. Additionally, our results are in line with previous research that also found that parks do have a cooling effect on urban areas [27,38,57,58].

This information is useful as it gives an idea of how the park's cooling effect changes over distance. If parks were planned better to maximize their cooling effects, taking into consideration the factors investigated in Sections 3.1–3.3, and we could plan proper buffer distances, then less parks would be necessary.

4. Conclusions

While previous studies have analyzed the impacts of various LULC categories on LST and PCI, our study utilizes a novel methodology incorporating XGBoost machine learning algorithm and state-of-the-art SHAP values for a more rounded analysis benefiting from SHAP's more unified approach to understanding complex model outputs. Unlike many prediction models, SHAP value outputs are not focused on quantifying how much change the input variables have caused to the output but focus more specifically on how the inputs are impacting the output. In other words, it is not trying to answer, "how much" but "why". Furthermore, by using SHAP values, we can analyze the results at the individual park level as well as at a more aggregate level without compromising our model or results.

As LST and PCI are two factors that both have an impact on the urban heat island phenomena [35,36], this area of research is growing in importance. Furthermore, our results give extra reference to fill gaps in previous research that has focused more on rapidly growing and urbanizing cities that are denser (and densifying) and less on cities that are less dense but still rapidly growing, with large amounts of urban sprawl and suburbanization, such as Dallas, Texas.

In the end, the analysis of this study found that greener areas tend to have a higher impact on both lower LSTs and lower PCIs. This is in line with previous literature such as Du et al., 2017 [35] and Oliveira et al., 2011 [36] who studied Green-space Cooling Intensity (GCI) and Park Cooling Intensity (PCI), with us using PCI in our study. Other researchers have utilized PCI and remote sensing to study the effects of parks and green areas on lowering temperatures and found similar results that parks have a cooling affect [38,40,47,61]. Various case studies have also found the greening of cities as a significant source of UHI mitigation [5,25,27,33]. While our results are very much in line with previous studies, despite different methodologies and analytical tools, it is impactful to show that the cooling effect of green areas has a measurable impact in a range of urban areas from dense to sprawling.

When considering built-up areas, "misc. built-up" zones had a larger impact LST than PCI. On the other hand, "misc. open (green)" spaces had the highest impact on PCI and the second highest on LST. This is in line with previous research showing that artificial increases of temperatures in cities is happening in large part due to changes in the built-up environment [21–23,62].

Commercial and office zones, generally being more built-up, were found to have a bigger impact on LST than do residential zones; however, when considering PCI, they have practically the same impact. This result is both novel and intriguing as it shows the first divergence of LULC classifications affecting LST and PCI differently. In this specific case, we saw that commercial and office zones have a larger impact on the temperatures themselves than does residential zones, however, when looking at the impact they have on the surrounding park's ability to cool the buffer zones, they have relatively the same impact; these findings were in line with other research finding that residential as well as middle density are major contributors to UHI. This is an important finding when it comes to planning the zoning and placement of parks.

These results overall indicate that green space planning is crucial for both LST and PCI impacts, and better green space planning can help to mitigate urban heat island effects. Furthermore, our results show that LULC planning should consider currently existing parks in the planning area when attempting to mitigate urban heat island as the different categories affect the LST and PCI differently.

In terms of green space planning in the city of Dallas, a larger focus on planning near and within built-up areas is needed. Building within the areas can more efficiently lower the LST of the zone, however, there are acting forces against the PCI that limit its reach. With that, building parks towards the outside of the built-up zones and nearer to other existing parks and green spaces can help increase the PCI of the new park and extend its reach.

In other words, for adequate green space planning to mitigate UHI, it is necessary to focus not only on building parks in areas that are built-up (which predominantly lowers LST in just that area) but also taking into consideration existing parks and green spaces that can have positive effects on nearby parks, increasing and extending the PCI of the new park (which lowers the temperatures of both the area and surrounding areas). Although creating new parks in empty/vacant lots will anyway have heat reduction effects, local planners are recommended to develop future parks in appropriate places, where they can maximize the UHI reduction impacts. A local comprehensive, green space plan, as well as several land and tree preservation codes should also be adapted by recognizing the differential effects of parks and green spaces to nearby areas and conduct more systematic suitability analysis for finding potential park locations.

Furthermore, our study was able to conclude that the utilization of machine learning boosting algorithms, in our case XGBoost, to generate SHAP values, is indeed a viable method to study the impact of different factors, with us focusing on LULC classification, LST and PCI, two factors found to correlate with UHI. This novel method allows for quicker analyses over larger areas of studies as much of the previous research, obtaining the same results, focused on case studies or smaller level analyses [25–27].

Limitations

While this study found several noteworthy policy implications that need to be concerned during the future park development, some limitations still exist, which need to be further concerned in the future study.

First, the main limitations of this study are centered on data availability. For our LULC data, we had to use data produced at the county level for the city of Dallas in 2015. While our results are in line with previous research, a more micro level data input could have potentially yielded more detailed results. As well, more up-to-date data and data over more frequent periods could have contributed to a better study of the city.

However, the 2015 dataset was the most up-to-date available data at the time of this study. The LULC data is produced at 5-year intervals, making the data a challenge to perform longitudinal studies on as the classifications can change significantly over that time. To help solve this data issue, future research could focus on more novel ways to collect LULC data, potentially utilizing satellite images and machine learning techniques.

Again, while this study did align with previous study's results, another data limitation is the lack of micro level weather data to add as factors in our model. As mentioned beforehand, microclimate and meteorological data has a potentially large impact on the temperature of the study area. There is an innate complexity in local climates and related environmental conditions. The measurement of air temperature is limited by the monitoring system, including pieces of equipment, experts, methods, and related factors. The accuracy of data collection is a key factor, and it is difficult to conduct ideal UHI research on a wide range of space-time scales. These data sources are, however, useful in understanding the generalized sources of data studied. To help lessen the effects of missing these variables, our final LST calculations are based on an average of multiple days over the same month in the study area to help give a more rounded estimation of the temperatures.

Future studies could benefit from incorporating more advanced data detection methods such as utilizing satellite images and machine learning algorithms to detect and classify the various LULC categories or potentially use more advanced models to assist with microclimate data interpolation in areas with less dense spreads of weather monitoring stations. While these would make for excellent research, they are beyond the scope of this current paper, which is focused on utilizing boosting algorithms and SHAP values to build impact assessment models.

Additionally, future studies could incorporate other methods and expand on previous studies such as Li, et al. [63], which utilized machine learning techniques to predict land use demand and zoning with a focus on urban-industrial zones. By expanding the study, it would be possible to better predict potential future impacts of zoning and thus be able to better plan mitigation of UHI.

Another limitation, which comes with most studies utilizing machine learning, is the limitations around model creation, tuning, and processing. While we have taken all precautions possible to build the best models with the least error and bias, there will always be other considerations that could have been taken to improve the models. To help mitigate this limitation, our approach utilized Bayesian Optimization for our parameter tuning and followed a basic cost–benefit curve comparing accuracy to processing power required.

Author Contributions: Conceptualization, D.M. and H.W.K.; data collection, D.M. and H.W.K.; methodology, D.M. and H.W.K.; formal analysis, D.M.; investigation, D.M.; project administration, D.M., J.L. and H.W.K.; writing—original draft, D.M.; writing—review & editing, J.L. and H.W.K. All authors have read and agreed to the published version of the manuscript.

Funding: This research was supported by Incheon National University Research Grant in 2021, grant number 2021-0064.

Conflicts of Interest: The authors declare no conflict of interest.

Appendix A

Table A1. Original 33 Land Use Land Cover classifications and 11 groupings.

Land Use/Land Cover Category	Grouping
Water	Water
Transit	Transit
Mobile home	Residential
Group quarters	Residential
Single family	Residential
Multi family	Residential
Parks recreation	Parks
Parking	Parking
Mixed use	Mixed-use
Landfill	Misc open (non-green)
Under construction	Misc open (non-green)
Cemeteries	Misc open (non-green)
Farmland	Misc open (non-green)
Improved acreage	Misc open (non-green)
Cultivated crops	Misc open (non-green)
Barren land	Misc open (non-green)
Residential acreage	Misc open (non-green)
Vacant	Misc open (non-green)
Ranch land	Misc open (non-green)
Communication	Misc built-up
Large stadium	Misc built-up
Airport	Misc built-up
Runway	Misc built-up
Utilities	Misc built-up
Railroad	Misc built-up
Industrial	Industrial
Flood control	Flood control
Office	Commercial/office
Hotel motel	Commercial/office
Retail	Commercial/office
Institutional semi public	Commercial/office
Commercial	Commercial/office
Education	Commercial/office

References

1. Dietz, M.E. Low impact development practices: A review of current research and recommendations for future directions. *Water Air Soil Pollut.* **2007**, *186*, 351–363. [CrossRef]
2. Kim, H.W.; Park, Y. Urban green infrastructure and local flooding: The impact of landscape patterns on peak runoff in four Texas MSAs. *Appl. Geogr.* **2016**, *77*, 72–81. [CrossRef]
3. Pugh, T.A.M.; Mackenzie, A.R.; Whyatt, J.D.; Hewitt, C.N. Effectiveness of green infrastructure for improvement of air quality in urban street canyons. *Environ. Sci. Technol.* **2012**, *46*, 7692–7699. [CrossRef] [PubMed]

4. Turner, K.; Lefler, L.; Freedman, B. Plant communities of selected urbanized areas of Halifax, Nova Scotia, Canada. *Landsc. Urban Plan.* **2005**, *71*, 191–206. [CrossRef]

5. Bowler, D.E.; Buyung-Ali, L.; Knight, T.M.; Pullin, A.S. Urban greening to cool towns and cities: A systematic review of the empirical evidence. *Landsc. Urban Plan.* **2010**, *97*, 147–155. [CrossRef]

6. Oke, T.R. City size and the urban heat island. *Atmos. Environ.* **1973**, *7*, 769–779. [CrossRef]

7. Stewart, I.D.; Oke, T.R.; Krayenhoff, E.S. Evaluation of the 'local climate zone' scheme using temperature observations and model simulations. *Int. J. Climatol.* **2014**, *34*, 1062–1080. [CrossRef]

8. Deilami, K.; Kamruzzaman, M.; Liu, Y. Urban heat island effect: A systematic review of spatio-temporal factors, data, methods, and mitigation measures. *ITC J.* **2018**, *67*, 30–42. [CrossRef]

9. Spence, M.; Annez, P.; Buckley, R. *Urbanization and Growth*; World Bank: Washington, DC, USA, 2008.

10. United States Environmental Protection Agency (US EPA). *OAR Heat Island Effect*; United States Environmental Protection Agency (US EPA): Washington, DC, USA, 2014.

11. Abutaleb, K.; Ngie, A.; Darwish, A.; Ahmed, M.; Arafat, S.; Ahmed, F. Assessment of urban heat island using remotely sensed imagery over greater Cairo, Egypt. *Adv. Remote Sens.* **2015**, *4*, 35–47. [CrossRef]

12. Yang, J.; Ren, J.; Sun, D.; Xiao, X.; Xia, J.C.; Jin, C.; Li, X. Understanding land surface temperature impact factors based on local climate zones. *Sustain. Cities Soc.* **2021**, *69*, 102818. [CrossRef]

13. Oliver, J.E.; Oke, T.R. Boundary Layer Climates. *Geogr. Rev.* **1979**, *69*, 486. [CrossRef]

14. Fonseka, H.P.U.; Zhang, H.; Sun, Y.; Su, H.; Lin, H.; Lin, Y. Urbanization and its impacts on land surface temperature in Colombo Metropolitan Area, Sri Lanka, from 1988 to 2016. *Remote Sens. Basel* **2019**, *11*, 957. [CrossRef]

15. Efe, S.I.; Eyefia, O.A. Urban Warming in Benin City, Nigeria. *Atmos. Clim. Sci.* **2014**, *4*, 241–252. [CrossRef]

16. Lee, H.; Kim, H.W. Examining the effects of green infrastructure configurations and patterns on urban thermal environment in Incheon, South Korea. *J. Korean Reg. Dev.* **2020**, *32*, 67–94.

17. Oke, T.R. The energetic basis of the urban heat island. *Q. J. R. Meteorol. Soc.* **1982**, *108*, 1–24. [CrossRef]

18. Tapper, N.J. Urban Influences on Boundary-Layer Temperature and Humidity - Results from Christchurch, NewZealand. *Atmos. Environ. Part B Urban Atmos.* **1990**, *24*, 19–27. [CrossRef]

19. Paterson, D.A.; Apelt, C.J. Simulation of Wind Flow around 3-Dimensional Buildings. *Build. Environ.* **1989**, *24*, 39–50. [CrossRef]

20. Winguth, A.M.E.; Kelp, B. The urban heat island of the north-central Texas region and its relation to the 2011 severe Texas drought. *J. Appl. Meteorol. Climatol.* **2013**, *52*, 2418–2433. [CrossRef]

21. Erell, E.; Pearlmutter, D.; Williamson, T. *Urban Microclimate: Designing the Spaces between Buildings*; Earthscan: London, UK, 2011.

22. Santamouris, M. Heat island research in Europe: The state of the art. *Adv. Build. Energy Res.* **2007**, *1*, 123–150. [CrossRef]

23. Yuan, F.; Bauer, M.E. Comparison of impervious surface area and normalized difference vegetation index as indicators of surface urban heat island effects in Landsat imagery. *Remote Sens. Environ.* **2007**, *106*, 375–386. [CrossRef]

24. Sharifi, E.; Lehmann, S. Case study of central Sydney. *J. Urban Environ. Eng.* **2015**, *9*, 3–11. [CrossRef]

25. Doick, K.J.; Peace, A.; Hutchings, T.R. The role of one large greenspace in mitigating London's nocturnal urban heat island. *Sci. Total Environ.* **2014**, *493*, 662–671. [CrossRef]

26. Kawashima, S.; Ishida, T.; Minomura, M.; Miwa, T. Relations between surface temperature and air temperature on a local scale during winter nights. *J. Appl. Meteorol.* **2000**, *39*, 1570–1579. [CrossRef]

27. Yan, H.; Wu, F.; Dong, L. Influence of a large urban park on the local urban thermal environment. *Sci. Total Environ.* **2018**, *622–623*, 882–891. [CrossRef]

28. Yu, Z.; Xu, S.; Zhang, Y.; Jørgensen, G.; Vejre, H. Strong contributions of local background climate to the cooling effect of urban green vegetation. *Sci. Rep.* **2018**, *8*, 6798. [CrossRef]

29. Yu, Z.; Guo, X.; Zeng, Y.; Koga, M.; Vejre, H. Variations in land surface temperature and cooling efficiency of green space in rapid urbanization: The case of Fuzhou city, China. *Urban For. Urban Green.* **2018**, *29*, 113–121. [CrossRef]

30. Chatterjee, S.; Khan, A.; Dinda, A.; Mithun, S.; Khatun, R.; Akbari, H.; Kusaka, H.; Mitra, C.; Bhatti, S.S.; Van Doan, Q.; et al. Simulating micro-scale thermal interactions in different building environments for mitigating urban heat islands. *Sci. Total Environ.* **2019**, *663*, 610–631. [CrossRef]

31. Lindén, J.; Fonti, P.; Esper, J. Temporal variations in microclimate cooling induced by urban trees in Mainz, Germany. *Urban For. Urban Green.* **2016**, *20*, 198–209. [CrossRef]

32. Yahia, M.W.; Johansson, E.; Thorsson, S.; Lindberg, F.; Rasmussen, M.I. Effect of urban design on microclimate and thermal comfort outdoors in warm-humid Dar es Salaam, Tanzania. *Int. J. Biometeorol.* **2018**, *62*, 373–385. [CrossRef]

33. Ng, E.; Chen, L.; Wang, Y.; Yuan, C. A study on the cooling effects of greening in a high-density city: An experience from Hong Kong. *Build. Environ.* **2012**, *47*, 256–271. [CrossRef]

34. Rasul, A.; Balzter, H.; Smith, C.; Remedios, J.; Adamu, B.; Sobrino, J.; Srivanit, M.; Weng, Q. A review on remote sensing of urban heat and cool islands. *Land Basel* **2017**, *6*, 38. [CrossRef]

35. Du, H.; Cai, W.; Xu, Y.; Wang, Z.; Wang, Y.; Cai, Y. Quantifying the cool island effects of urban green spaces using remote sensing Data. *Urban For. Urban Green.* **2017**, *27*, 24–31. [CrossRef]

36. Oliveira, S.; Andrade, H.; Vaz, T. The cooling effect of green spaces as a contribution to the mitigation of urban heat: A case study in Lisbon. *Build. Environ.* **2011**, *46*, 2186–2194. [CrossRef]

37. Cao, X.; Onishi, A.; Chen, J.; Imura, H. Quantifying the cool island intensity of urban parks using ASTER and IKONOS data. *Landsc. Urban Plan.* **2010**, *96*, 224–231. [CrossRef]

38. Lin, W.; Yu, T.; Chang, X.; Wu, W.; Zhang, Y. Calculating cooling extents of green parks using remote sensing: Method and test. *Landsc. Urban Plan.* **2015**, *134*, 66–75. [CrossRef]

39. Xu, H. Analysis of impervious surface and its impact on urban heat environment using the normalized difference impervious surface index (NDISI). *Photogramm. Eng. Remote Sensing* **2010**, *76*, 557–565. [CrossRef]

40. Feyisa, G.L.; Dons, K.; Meilby, H. Efficiency of parks in mitigating urban heat island effect: An example from Addis Ababa. *Landsc. Urban Plan.* **2014**, *123*, 87–95. [CrossRef]

41. Kanda, M.; Moriizumi, T. Momentum and heat transfer over urban-like surfaces. *Boundary Layer Meteorol.* **2009**, *131*, 385–401. [CrossRef]

42. Kotthaus, S.; Grimmond, C.S.B. Energy exchange in a dense urban environment—Part I: Temporal variability of long-term observations in central London. *Urban Clim.* **2014**, *10*, 261–280. [CrossRef]

43. Monteith, J.L.; Unsworth, M.H. *Principles of Environmental Physics: Plants, Animals, and the Atmosphere*, 4th ed.; Elsevier/Academic Press: Amsterdam, The Netherlands, 2013.

44. Efron, B.; Morris, C. Stein's Paradox in Statistics. *Sci. Am.* **1977**, *236*, 119–127. [CrossRef]

45. Sultana, S.; Satyanarayana, A.N.V. Impact of urbanisation on urban heat island intensity during summer and winter over Indian metropolitan cities. *Environ. Monit. Assess.* **2020**, *191*, 789. [CrossRef] [PubMed]

46. NCTCOG. 2015 Land Use. Available online: https://data-nctcoggis.opendata.arcgis.com/datasets/2015-land-use/explore (accessed on 5 May 2021).

47. Chibuike, E.M.; Ibukun, A.O.; Abbas, A.; Kunda, J.J. Assessment of green parks cooling effect on Abuja urban microclimate using geospatial techniques. *Remote Sens. Appl. Soc. Env.* **2018**, *11*, 11–21. [CrossRef]

48. Annerstedt, M.; Ostergren, P.-O.; Björk, J.; Grahn, P.; Skärbäck, E.; Währborg, P. Green qualities in the neighbourhood and mental health—Results from a longitudinal cohort study in Southern Sweden. *BMC Public Health* **2012**, *12*, 337. [CrossRef]

49. Giles-Corti, B.; Broomhall, M.H.; Knuiman, M.; Collins, C.; Douglas, K.; Ng, K.; Lange, A.; Donovan, R.J. Increasing walking: How important is distance to, attractiveness, and size of public open space? *Am. J. Prev. Med.* **2005**, *28*, 169–176. [CrossRef]

50. Schipperijn, J.; Ekholm, O.; Stigsdotter, U.K.; Toftager, M.; Bentsen, P.; Kamper-Jørgensen, F.; Randrup, T.B. Factors influencing the use of green space: Results from a Danish national representative survey. *Landsc. Urban Plan.* **2010**, *95*, 130–137. [CrossRef]

51. Parastatidis, D.; Mitraka, Z.; Chrysoulakis, N.; Abrams, M. Online global land surface temperature estimation from Landsat. *Remote Sens. Basel* **2017**, *9*, 1208. [CrossRef]

52. Weng, Q.; Lu, D.; Schubring, J. Estimation of land surface temperature–vegetation abundance relationship for urban heat island studies. *Remote Sens. Environ.* **2004**, *89*, 467–483. [CrossRef]

53. Camilloni, I.; Barrucand, M. Temporal variability of the Buenos Aires, Argentina, urban heat island. *Theor. Appl. Climatol.* **2012**, *107*, 47–58. [CrossRef]

54. Li, Z.-L.; Tang, B.-H.; Wu, H.; Ren, H.; Yan, G.; Wan, Z.; Trigo, I.F.; Sobrino, J.A. Satellite-derived land surface temperature: Current status and perspectives. *Remote Sens. Environ.* **2013**, *131*, 14–37. [CrossRef]

55. Peng, S.; Piao, S.; Ciais, P.; Friedlingstein, P.; Ottle, C.; Bréon, F.-M.; Nan, H.; Zhou, L.; Myneni, R.B. Surface urban heat island across 419 global big cities. *Environ. Sci. Technol.* **2012**, *46*, 696–703. [CrossRef]

56. Rogan, J.; Ziemer, M.; Martin, D.; Ratick, S.; Cuba, N.; DeLauer, V. The impact of tree cover loss on land surface temperature: A case study of central Massachusetts using Landsat Thematic Mapper thermal data. *Appl. Geogr.* **2013**, *45*, 49–57. [CrossRef]

57. Liu, H.; Weng, Q. Seasonal variations in the relationship between landscape pattern and land surface temperature in Indianapolis, USA. *Environ. Monit. Assess.* **2008**, *144*, 199–219. [CrossRef]

58. Spronken-Smith, R.A.; Oke, T.R. The thermal regime of urban parks in two cities with different summer climates. *Int. J. Remote Sens.* **1998**, *19*, 2085–2104. [CrossRef]

59. Chen, T.; Guestrin, C. XGBoost: A Scalable Tree Boosting System. *arXiv* **2016**, arXiv:1603.02754.

60. Lundberg, S.; Lee, S.-I. A unified approach to interpreting model predictions. *arXiv* **2017**, arXiv:1705.07874.

61. Qiu, K.; Jia, B. The roles of landscape both inside the park and the surroundings in park cooling effect. *Sustain. Cities Soc.* **2020**, *52*, 101864. [CrossRef]

62. Geros, V.; Santamouris, M.; Amourgis, S.; Medved, S.; Milford, E.; Robinson, G.; Steemers, K.; Karatasou, S. A distant-learning training module on the environmental design of urban buildings. *Renew. Energy* **2006**, *31*, 2447–2459. [CrossRef]

63. Li, C.; Gao, X.; Wu, J.; Wu, K. Demand prediction and regulation zoning of urban-industrial land: Evidence from Beijing-Tianjin-Hebei Urban Agglomeration, China. *Environ. Monit. Assess.* **2019**, *191*, 412. [CrossRef] [PubMed]

 *sustainability*

Article

Literature Review Reveals a Global Access Inequity to Urban Green Spaces

Yan Sun [1], Somidh Saha [2,3], Heike Tost [4], Xiangqi Kong [1] and Chengyang Xu [1,*]

1   The Key Laboratory for Silviculture and Conservation of Ministry of Education, Key Laboratory for Silviculture and Forest Ecosystem of State Forestry and Grassland Administration, Research Center for Urban Forestry, Beijing Forestry University, Beijing 100083, China; yanearth@bjfu.edu.cn (Y.S.); xiangqi333@bjfu.edu.cn (X.K.)
2   Research Group Sylvanus, Institute for Technology Assessment and Systems Analysis, Karlsruhe Institute of Technology, Karlstr. 11, 76133 Karlsruhe, Germany; somidh.saha@kit.edu
3   Chair of Silviculture, Institute of Forest Sciences, Faculty of Environment and Natural Resources, University of Freiburg, 79106 Freiburg, Germany
4   Department of Psychiatry and Psychotherapy, Central Institute of Mental Health, Medical Faculty Mannheim, University of Heidelberg, J5, 68159 Mannheim, Germany; heike.tost@zi-mannheim.de
*   Correspondence: cyxu@bjfu.edu.cn; Tel.: +86-10-6233-7082

Abstract: Differences in the accessibility to urban resources between different racial and socioeconomic groups have exerted pressure on effective planning and management for sustainable city development. However, few studies have examined the multiple factors that may influence the mitigation of urban green spaces (UGS) inequity. This study reports the results of a systematic mapping of access inequity research through correspondence analysis (CA) to reveal critical trends, knowledge gaps, and clusters based on a sample of 49 empirical studies screened from 563 selected papers. Our findings suggest that although the scale of cities with UGS access inequity varies between countries, large cities (more than 1,000,000 population), especially in low- and middle-income countries (LMICs), are particularly affected. Moreover, the number of cities in which high socioeconomic status (high-SES) groups (e.g., young, rich, or employed) are at an advantage concerning access to UGS is substantially higher than the number of cities showing better accessibility for low-SES groups. Across the reviewed papers, analyses on mitigating interventions are sparse, and among the few studies that touch upon this, we found different central issues in local mitigating strategies between high-income countries (HICs) and LMICs. An explanatory framework is offered, explaining the interaction between UGS access inequity and local mitigating measures.

Keywords: access inequity; systematic mapping; empirical studies; city scale; inequity mitigation

Citation: Sun, Y.; Saha, S.; Tost, H.; Kong, X.; Xu, C. Literature Review Reveals a Global Access Inequity to Urban Green Spaces. *Sustainability* **2022**, 14, 1062. https://doi.org/10.3390/su14031062

Academic Editor: Hyun-Woo Kim

Received: 30 November 2021
Accepted: 13 January 2022
Published: 18 January 2022

Publisher's Note: MDPI stays neutral with regard to jurisdictional claims in published maps and institutional affiliations.

1. Introduction

Over half of the world's population now live in urban areas, and this proportion is projected to increase to about 75% by 2050 [1]. Urban green spaces (UGS), as the critical connection between the natural environment and human beings in cities, can deliver a variety of ecosystem services that increase both the life quality and resilience of urban dwellers [2–4]. Moreover, UGS have been linked to increased physical and mental well-being, reduced psychological morbidity, and enhanced social cohesion [5,6]. Urban residents might be particularly vulnerable to the disparity between green, blue, and built infrastructures under the complicated circumstances of the multiple interacting ecological, social, and technological drivers of urban expansion. At the same time, disparities in access to UGS have been found among groups with different socioeconomic statuses (SES, often measured by income, education, and occupation), and this inequity has been investigated in different geographical settings. Access to UGS has been further conceptualized by Rigolon [7] in the light of three aspects—proximity, quantity, and quality—to compare empirical studies in high-income countries (HICs), and the geographical differences between inner-city areas (where UGS are relatively sparse and low-income residents are frequent) and suburbs might also contribute to access inequity. A recent comparative study of 290 Chinese cities revealed ubiquitous differences in UGS exposure between old and newly urbanized areas [8]. Rapidly urbanizing cities have been confronted with the sustainability issue of urban green system development and adding new parks might cause the increase in nearby housing prices and subsequent neighboring gentrification [9,10]. Although this inequity has been quantified and mapped, the underlying mechanisms of UGS access inequity and potential mitigation measures are not fully understood. Economic development and urban afforestation-related policies might exert divergent impacts on UGS distributions [11]. A review of the local contexts moderating UGS access inequality in various cities across different countries and continents may provide critical insights in this field of research.

The past two decades have seen increasing concerns related to UGS access and associated inequity. Empirical studies have been published in journals from various disciplines, including land use, urban planning, sustainable science, public health, and environmental science [12]. Research related to access inequity has focused on the disparity among groups based on income, age, gender, education, family structure, employment, race, and ethnicity. Previous research attempted to identify the critical areas of access inequity in densely populated areas [13] of a city, or the newly developed areas in a city's rural or suburban areas [14]. Moreover, quantitative models have been developed and applied to reveal unfairness in the provision of UGS facilities, especially the quantity and proximity aspects of accessibility [15]. Another significant concern that accompanies the inequity is its consequences for social interaction and health inequalities [16]. To improve human health and well-being, various forms of UGS in different communities of a city require elaborate design, planning, and management [17,18]. Therefore, in urban areas, the associations between access inequity to UGS and human health inequities at both local and regional scales are important. Furthermore, more epidemiological evidence that combines spatially explicit urban environment data with socioeconomic conditions and individual health data is needed to dissect the individual and societal effects of inequity in the proximity, quantity, and quality of UGS. Mitigating inequity in access to green spaces in urban environments could deliver a triple win, including strengthened social cohesion, improved health outcomes, and sustainable urban ecosystems. Nevertheless, few systematic reviews of empirical studies on this topic have been published, and a synthesis of the findings and remaining gaps in knowledge is lacking.

The main objectives of this study were to reveal the relationship between city-scale and access inequity from global empirical studies, examine the multiple factors that may influence the mitigation of UGS inequity, and provide a systematic map to identify the gaps and clusters in research on inequity in access to UGS. We expect that our effort to collate evidence covering the field's breadth will stimulate future research on UGS access inequity.

2. Methods

2.1. Literature Search

To conduct a systematic review of the literature on spatial accessibility to UGS in the context of inequity, we searched the databases of the Web of Science focused on empirical studies from 2000 to 2020. We focused on empirical studies using the method of equity mapping, which was developed in the late 1990s [7], and thus, we did not search studies published before 2000. We searched papers in peer-reviewed journals using the following multiple terms: (greenspace * or green space * or open space * or urban park * or park * or playground *) AND (access * or "accessibility" or "spatial distribution" or "provision" or equit * or inequ * or disparit * or "socioeconomic" or "socioeconomic" or "income"). Studies were identified and screened with the following criteria:

(i)   English language and full text available: Only studies published in English and studies with available full-text versions were included. Comments, conference abstracts, book chapters, and reviews were excluded.

(ii)   Empirical research: Studies that demonstrated accessibility using spatial data with a quantitative measure under the scope of proximity, quantity, or quality were included. Purely descriptive studies were excluded.

(iii)   Index of inequity: Studies that performed inequity analysis were required to include a quantitative measure related to socioeconomic or ethnic status.

(iv)   Study objects: Studies that focused on the intra-city inequity pattern were included. Studies that targeted more than one city but performed intra-city quantitative analysis for each city were also included. Multi-city studies that solely assessed inter-city differences were excluded.

The literature search initially identified a total of 563 articles, of which 118 articles were further screened in full text (Figure 1) after the title and abstract screening. In total, 49 studies were included in the review (Table 1).

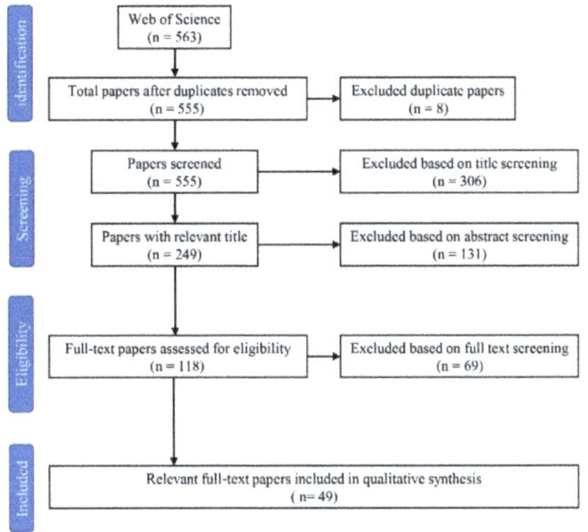

Figure 1. The conceptual diagram for the literature search process.

Table 1. Articles selected for systematic mapping.

Author(s) and Date		
Abercrombie et al. [19]	Almohamad et al. [20]	Arroyo-Johnson et al. [21]
Astell-Burt et al. [22]	Bahrini et al. [23]	Barbosa et al. [24]
Bruton et al. [25]	Chen et al. [26]	Comber et al. [27]
Cradock et al. [28]	Dadvand et al. [29]	Dai [15]
de Mola et al. [30]	Engelberg et al. [31]	Feng et al. [32]
Gu et al. [33]	Guo et al. [34]	He et al. [35]
Hoffimann et al. [36]	Iraegui et al. [37]	Jenkins et al. [38]
Kabisch & Haase [13]	Kamel et al. [39]	Knapp et al. [40]
La Rosa et al. [41]	Lara-Valencia & Garcia-Perez [42]	Lara-Valencia & Garcia-Perez [43]
Lin et al. [44]	Manta et al. [45]	Nero [46]
Park et al. [47]	Rahman & Zhang [48]	Reyes et al. [49]
Sathyakumar et al. [50]	Schule et al. [51]	Shen et al. [52]
Sugiyama et al. [53]	Tan et al. [54]	Tan & Samsudin [55]
Tian et al. [56]	Tu et al. [57]	Wei [58]
Weiss et al. [59]	Wende et al. [60]	Xiao et al. [14]
Xu et al. [61]	Yang et al. [62]	Zhang et al. [63]
Zhou & Kim [64]		

2.2. Coding Methods to Perform Research Synthesis

To analyze the articles that met the review criteria, we used research feature variables to code the studies for synthesis. The codes were key variables for correspondence analysis (CA), including the type of access to UGS, measures of SES, findings for inequity, the country characteristics of the studied city, the population of the studied city, the spatial and temporal analysis scale, and the intervention analysis (Table 2). These variables were adopted to find critical trends, research gaps, and clusters in access inequity.

Table 2. Coded variables for correspondence analysis (CA).

Code Variables	Categories	Definition
Type of access studied	Proximity	Distance to UGS
	Quantity	Amount or coverage of UGS within given areas
	Quality	Multi-dimensional features of UGS (e.g., amenities, safety, biodiversity)
	Multiple	Two or more aspects of access were studied
Measures of SES	Single	Only one measure of SES
	Multiple	Two or more measures of SES
Inequity	Yes/No	Studies defined as "Yes" explicitly stated that they found differences in access to UGS among groups of different SES status
Country characteristics	HICs	The city (or cities) studied in the article is in a high-income country
	LMICs	The city (or cities) studied in the article is in a low- or middle-income country
City size	More than 1,000,000	The population of the studied city; the multi-city studies containing one city of more than 1,000,000 population were also coded into this category
	Less than 1,000,000	The population of the studied city (or cities)
Temporal scale	Cross-sectional	The empirical analyses were based on data at a single point in time
	Longitudinal	The empirical analyses were based on multiple-year data
Spatial scale	Single scale	Results were analyzed at a single spatial resolution
	Multiple scales	Results were analyzed at different spatial resolutions
Interventions	Yes/No	Studies defined as "Yes" mentioned the influence of local policy, planning, or initiatives on inequity

Note: HICs: high-income countries; LMICs: low- and middle-income countries; UGS: urban green spaces.

2.3. Evolution of Quantitative Models for Estimating UGS Accessibility

Various methods were applied to quantify the accessibility of UGS, and hence, we summarized the changes in those quantitative approaches in the articles analyzed in this study. We mapped the evolution of quantitative models on a time axis. Furthermore, we categorized the studies according to three typical aspects (i.e., quantity, proximity, and quality) of access to UGS [7] and listed the years of publication and the corresponding number of those publications (Supplementary, Figure S1).

2.4. Spatial Distribution of the Studied Cities and Access Inequity

To further illustrate the distribution patterns of studied cities, we analyzed the empirical studies according to their geographical areas. The single analysis unit of a country might not imply the varied spatial distribution patterns of the city under examination; we classified the studied cities by the population scale to which they belong. Accordingly, we checked each paper and recorded the specific city, county, or district that had been its focus on and then mapped the distribution using the vector data from the Database of Global Administrative Areas (GADM) for administrative subdivisions (https://gadm.org/, accessed on 17 February 2021).

The SES of urban residents can be identified from income, education, gender, age, housing price, employment, as well as ethnic/racial factors. Previous studies from HICs tended to emphasize the inequity of UGS accessibility among different ethnic groups (e.g., white or Latino), and most studies from LMICs did not analyze the variations of accessibility from the aspect of ethnic/racial groups. We gathered the advantage or disadvantage patterns from the 49 included articles and checked the inequity pattern of access to UGS.

2.5. Visualizing the Synthesis Analysis with Correspondence Analysis

Systematic mapping has been accepted as an effective method to describe and identify the characteristics of evidence synthesis. Given that the objectives of the review were not to answer a specific question about access inequity occurrence, but to present a broad synthesis of the access inequity research field to identify future study priorities, we did not choose meta-analysis but adopted CA as a combination of quantitative and qualitative measurement [65]. CA could facilitate the visualization of multiple features from reviewed studies in one figure through a display of a few dimensions. This method has been widely used in social sciences [66] and is now increasingly adopted in urban ecology [67,68], using novel data visualization methods to translate the evidence into a vital message for policymakers [69,70].

In this study, we characterized the individual empirical study using the nominal variables in Table 2 to link studies together and generated frequencies and two-way contingency tables. Mathematically, CA provides the spatial representation of similarity, similar to the principal component analysis, based on the chi-squared distance. Accordingly, we used CA to study the similarities between those empirical studies and examined whether some of those categories were closely related to each other. CA could visually display the rows and columns of the two-way contingency tables and describe the relationships between the categories of each variable. The CA biplots for the distribution pattern of variables could be used as an objective and effective method to demonstrate their interdependence. Specifically, a higher proximity of categories indicates a higher correspondence, and the same domain of the categories' location according to the axes means a positive clustering, which could elucidate the key trends of the research field. To conduct CA, in this study, we used the R package 'FactoMineR' [71] to create the CA plots [6].

3. Results

3.1. The Global Distribution Patterns of City-Scale and Access Inequity

Table 3 presents the spatial distribution of the studied cities. We found very few studies in large areas of LMICs (e.g., Africa, South America, South Asia, Southeast Asia, eastern Europe, and Russia). Specifically, the majority of cities in these studies are on the scale of more than a population of 1,000,000 people, such as Munich, New York City, Sydney, and Beijing. It is noteworthy that studies in LMICs, especially China, predominantly emphasized large cities with a population of more than 5,000,000 people (blue circles). Comparatively, the distribution of studied cities in Europe and North America covered a broader range of city-scale.

Table 4 shows the access inequity of each studied city, and we found a global inequity in access to UGS. The number of cities where the high-SES groups were advantaged was substantially larger than that of cities showing better accessibility for the low-SES groups. However, we found that the proportion of cities demonstrating inequity in LMICs (84.2%) is slightly smaller than that in HICs (78.8%). Specifically, previous studies did not find significant inequity in accessibility to UGS for Tehran [23], Sheffield [24], Barcelona [29], Seattle [31] and Phoenix [43], Catania [41], and Beijing [57]. Those cities are distributed in three continents of Asia, Europe, and North America. By contrast, studies carried out for Sydney [22], New York City [59], and Shanghai [14] obtained the opposite conclusions and found the advantageous status of low-SES groups in terms of accessibility to UGS.

Table 3. The city-scale (population) distributions from the reviewed empirical studies.

City Scale	City Number	Location
5,000,000 and greater	19	Australia, Bangladesh, Chile, China, Columbia, Ghana, Iran, Mexico, Peru, Singapore, South Korea, US
1,000,000 to 5,000,000	16	Argentina, Australia, Bolivia, Canada, Germany, India, Japan, Spain, Syria, US
500,000 to 1,000,000	6	Brazil, Mexico, UK, US
250,000 to 500,000	8	Italy, Portugal, UK, US
100,000 to 250,000	3	US

Table 4. The distribution of access inequity from the reviewed empirical studies.

Access Inequity	City Number	Location
Found	42	Argentina, Australia, Bangladesh, Bolivia, Brazil, Canada, Chile, China, Columbia, Germany, Ghana, Japan, India, Mexico, Peru, Portugal, Singapore, South Korea, Syria, UK, US
Not Found	10	China, Australia, Iran, Italy, Spain, UK, US

Those three cities are distributed in Asia, North America, and Oceania. There has been more than one study carried out on New York City, and the conclusion of inequity or equity is not absolute due to the research approach and data availability. Equitably distributed across different social groups might be the ideal goal of policymakers and practitioners. Xiao et al. [14] used the spatial clustering method to assess the social equity of accessibility to urban parks and found that two typical vulnerable groups (i.e., laid-off workers and rural migrants) are more favored over more affluent residents in Shanghai. They pointed out that this phenomenon might be attributed to the local municipal endeavors in ensuring the socially equitable access to parks, and it is noteworthy that the authors emphasized the critical influence of the planning legacy of China's socialist era. The study by Astell-Burt et al. [22] adopted the negative binomial and logit regression model to examine the inequitable distribution through a cross-city comparison and found the varied magnitude between cities and the intra regions within a city. They inferred that because of urban sprawl, those suburbs distant from business centers could have cheaper land and more green space.

3.2. Interventions for Access Inequity Mitigation

Figure 2 shows the major factors of SES in those 49 empirical articles on access inequity. The income and age traits of the residents might be the most important aspects to consider when mitigating the inequity in access to UGS. Meanwhile, a few studies focused on gender, employment, or/and family structure traits. While a local, targeted urban greening policy might result in the inequity pattern of access to parks, it can also act as an effective amelioration measure to address this issue [72,73]. However, across the reviewed papers, we found that analyses of mitigating interventions are sparse, and inadequate attention has been paid to the implementation phase and to evaluating the effects of specific local policies, planning, or practical initiatives to improve accessibility and mitigate inequity. Table 5 lists the interventions described in the reviewed articles. Research on large cities in the USA (i.e., Atlanta and Boston), and China (i.e., Beijing, Hangzhou, Nanjing, and Shanghai), place their policy influence analysis in their discussion sections. Scholars focused on initiatives related to local social structures (e.g., Singapore), long-term urban master plans, or current investment in parks. The policies from two empirical studies in the US [15,28] focused on redeveloping urban parks and paid attention to the parks' quality features (e.g., safety). Nevertheless, mitigation policies in the empirical studies in China shown in Table 5 mainly referred to the improvement of park areas and the walking time or distance to UGS, emphasizing the quantity and proximity aspects of accessibility. Notably, for Singapore, a typical multicultural city, social harmony might be the theme of urban green space planning and inequity mitigation interventions.

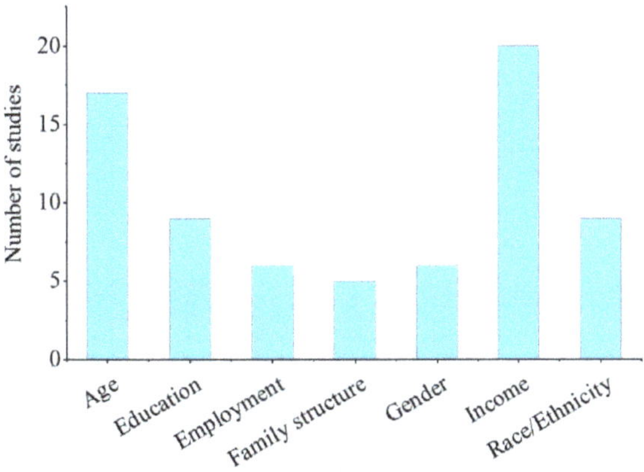

Figure 2. The major socioeconomic factors that influence access inequity.

3.3. Systematic Map of the Features of Access Inequity Research

To find critical trends, knowledge gaps, and clusters in access inequity research, we used key variables (Table 2) for CA to code the reviewed studies to produce a systematic map. The correspondence biplot (Figure 3) shows the diversity of approaches applied in access inequity research (blue text). The distance between any points or text gives a measure of their similarity. The plot displays the first two dimensions, which together explained 43.3% of the total variability (Table S1) in study features.

There is a small angle between the lines of the feature "proximity" and a lack of intervention analyses, as well as a small angle between studies measuring "quality" and adopting multiple measures of SES in the negative quadrant (Table S2). UGS "proximity" is the primary measure of accessibility (59% of the final studies for synthesis), and studies that analyzed access inequity under the scope of "quality" contributed the smallest proportion (Table S3). A total of 39 articles used multiple measures, and the proportion of adopting multiple measures of SES for "proximity", "quality", and "quantity" is 76.7%, 87.5%, and 85.2%, respectively (Table S4). The variable "proximity" displays the highest proportion (23.3%) for the research approach of using a single SES index. The proximity indicator weights the user's specified distance to measure the possibility of reaching an urban park, and the advance in access inequity research since 2000 has been accompanied by the increasing availability of spatial data and enhanced geographical information system (GIS) features, particularly the ArcGIS Network Analyst. The classic article of Dai (2011) [15] in the journal of Landscape and Urban Planning reported significantly poorer access to green spaces for neighborhoods with a higher proportion of African Americans in Atlanta. The author introduced a Gaussian-based two-step floating catchment (2SFCA) to assess the proximity, aiming to identify the key areas for intervention to address disparities. From the reviewed studies, the frequency of studies for "proximity" peaked in 2017 and kept being the dominant measure thereafter (Figure S1).

Table 5. Policies or interventions to mitigate access inequity described in the reviewed articles.

Author(s)	Study Location	Urban Green Space Planning and Inequity Mitigation Initiatives	Goals
Cradock et al., 2005 [28]	Boston, USA	Renovation by the Boston Parks and Recreation Department	Improve the safety of playgrounds
Dai, 2011 [15]	Atlanta, USA	Atlanta beltline redevelopment plan (Atlanta Development Authority, 2005)	Create 1200 acres of new or expanded parks; Improvements to over 700 acres of existing parks
Tan and Samsudin, 2017 [55]	Singapore	Ethnic Integration Policy	Maintain a racial mix quota in public housing estates and avoid forming racial enclaves in residential areas
Tu et al., 2018 [57]	Beijing, China	Urban Green Space System Planning (2004–2020)	Fund and build more than 100 public urban parks with a total area of 1700 hectares during 2005–2010
Wei, 2017 [58]	Hangzhou, China	Urban Green Space System Planning in Hangzhou The Green Space Planning in Hangzhou The Public Open Space Planning in Hangzhou	Recommend a 2 km distance for city parks and a 1–2 km for district parks with driving by private vehicles
Xiao et al., 2017 [14]	Shanghai, China	The 13th Five Year Plan's Public Green Space Special Plan (Ministry of Housing and Urban–Rural Development of the People's Republic of China, 2015)	Reduce the walking distance to public green space in the city proper to 500 m
Gu et al., 2017 [33]	Shanghai, China	Shanghai Master Plan (2015–2040)	Develop a new urban–rural parks system by 2040 to improve the accessibility of public green spaces
Zhang et al., 2019 [63]	Nanjing, China	The Planning of Nanjing City Parks (2017–2035)	Recommend that people enjoy a 10 min walk to the community-level park and a 20 min walk to the district-level park.

The variables "Without_Interventions_analyses" and "proximity" show similar trajectories in the CA biplot. There is either a weak or no correlation between "proximity" and "quantity" or "quality". UGS "quality" accounts for the least proportion in those studies that lack intervention analyses (Table S5). Both the proximity and quantity metrics ignore the inner characteristics of urban parks, and likely, that UGS quality is among the attributes that continually attracts nearby residents to visit those parks. A comparative study in a large US metropolitan area found significant differences in quality (i.e., location, care and maintenance, and entertainment value) between groups with different household income characteristics [38]. The influence of park quality on park usage might vary by gender, and the groups of active or female users could be significantly affected [40]. However, both of the above-mentioned empirical studies did not move forward to corresponding inequity-mitigating initiatives analysis.

There is a strong correlation between the studies in LMICs and studies in city sizes of more than 1,000,000 in the positive quadrant. Studies showing significant inequity of access were less distributed (32%) in LMICs (Table S6). However, the large angle between the lines of the feature "inequity" and cities in developing countries and the long distance between the two features indicate weak or no correlation. Although studies in LMICs have focused on large cities and found prevalent inequity, there has been a global inequity pattern despite the scenarios of LMICs or HICs.

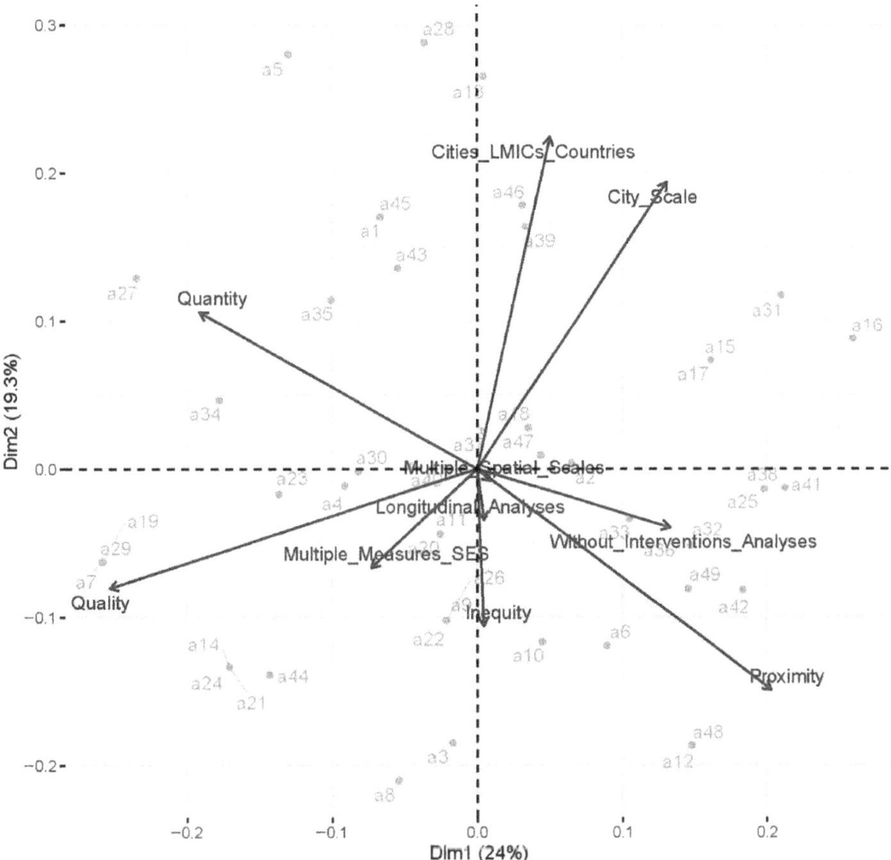

Figure 3. Biplot of the correspondence analysis with the 49 included studies. The studies are in gray points and characteristics (from Table 2) are in blue text. This symmetric plot shows the distribution pattern within the data. The percentage of explained variances of the first two dimensions are shown in the axes' labels. The distance between any points or text gives a measure of their similarity. Long lines of the text or points to the origin indicate a strong association. Small angles between two lines of text that point to the origin indicate associations. Lines with angles near 180 degrees present negative associations.

The variables of "Multiple_Spatial_Scales" and "Longitudinal_Analyses" showed a high level of similarity. Among the 49 reviewed articles for CA, only several articles were found to analyze the temporal dynamics of access inequity or analyzed inequity at different spatial scales (Tables S7 and S8). Wei [58] conducted a longitudinal study of Hangzhou, a megacity in China, and revealed the overall increase in park accessibility (i.e., proximity) from 2000 to 2010. This multi-temporal research was based on data acquisition from a variety of sources (i.e., statistical yearbooks, local government planning, and satellite images), but Wei found that neither access nor changes in access are significantly associated with the selected socioeconomic variables. Tan and Samsudin [55] explored the scale effects by comparing the park provisions and spatial equity from three different planning units (i.e., region, planning area, and subzones). They observed the pattern of stronger inequity with smaller scales in Singapore. To demonstrate the inequity variations, longitudinal and multi-spatial-scale research should address the challenge of overcoming socioeconomic and UGS distribution data availability and the selection of fitted statistical analysis models to reveal the mechanism of inequity dynamics.

4. Discussion

The global distribution map shows that studies have been concentrated in cities (blue circles) with more than 1,000,000 people, and large cities are predominantly affected by the inequity in access to UGS in LMICs. General interest in large cities might result from the fact that more data were available for large cities than for small cities in LMICs, as national governments tend to emphasize the sustainable development of capital cities and give priority to the budget for the census survey [74]. Population statistics data have been typically adopted to calculate accessibility when quantitative models (e.g., the 2SFCA were applied. The 2SFCA model was first proposed by Radke and Mu [75] to study spatial accessibility based on resource supplies and population demands. However, the administrative geographic units for the demographic data varied in many cases, and the spatialization method of population data to reduce the spatial mismatch between the urban park provision and population distribution pattern has not been investigated effectively. Recently, prognosis areas (sub-district level) in Berlin [13], US Census data by block group boundaries in Boston [28], and neighborhoods in Denver [76] were chosen as units from which demographic information was extracted. These studies assumed that the population is evenly distributed within the statistical unit, and therefore, they ignored the exceptionally high heterogeneity of population distribution in urban environments. The lack of spatialized population data of high resolution has been a limitation to improving inequity evaluation accuracy [77]. Tan et al. [54] used land cover type and demographic data to build a geographically weighted regression (GWR) model and generated the population grid map of 150 m resolution. They used the fine-scale population in their network analysis to obtain precise coverage data of different levels of accessibility. The input data of small-scale units for accessibility models can help diagnose the critical areas of spatial disparities to guide urban planning. Indeed, fine-scale building data could be incorporated into models for the grid map of demographic data [78], but those auxiliary data are often difficult to obtain.

The willingness of residents in large cities to visit urban forests might be influenced by the quality, due to various demands for recreation and relaxation [79]. The previous quantitative supply–demand models (e.g., the 2SFCA model), assuming that people are willing to visit nearby parks, might not work well if those parks are not aesthetically pleasing. Existing reviews have presented evidence suggesting that inequities of park quantity and quality exist in HICs and LMICs and that LMICs experienced a more consistent pattern of quality inequity than proximity and quantity [7,12]. Despite the recognized attraction of high park quality, present studies, in terms of urban park supply, have primarily focused on the presence/absence issues of urban parks rather than the intra-park characteristics. One of the most important reasons for this gap mentioned above is that researchers in this field may commonly rely on vector data (i.e., planning data for road and park boundaries). Admittedly, planning data could meet the research needs, to some extent, to demonstrate the inequality pattern of park coverage and proximity to urban residents. However, not enough is known about park quality, including vertical forest structure, vegetation type, landscape fragmentation, diversity, and complexity, all of which might influence the various ecosystem functions and services and could be effectively assessed using multi-source data [80,81]). The increasing availability of high-resolution, remote sensing data (e.g., images from QuickBird, Worldview, and SPOT satellites) and the LiDAR cloud data will allow for the spatially explicit evaluation of inequity concerning park quality. Future studies need to consider the increasing demand of urban residents to enjoy high-quality parks and explore the access inequity of parks at different quality levels in more detail. Future quantitative analyses should also focus on quantifying park-quality-related variables, such as forest structure, landscape diversity, and landscape fragmentation. More attention should be given to the integrated approach to strengthen the understanding of the driving mechanisms for park access inequity.

Of the existing access inequity studies, the majority are single-phase studies; hence, they lack a longitudinal analysis. A temporal comparative study can identify causal relationships between access inequity and its driving factors (e.g., socioeconomic, policy, or urban planning factors) and accordingly facilitate UGS planning optimization. The historical trajectories of infrastructure construction during the rapid urbanization process have shaped the spatial patterns of park access. Therefore, accessibility could be treated as a process with multiple dimensions rather than as an established state or outcome. The pioneering investigation of park access dynamics took place in Hangzhou, China [58]. This study demonstrated the predominant role of park investment on individual sub-districts to improve overall accessibility, but the research in Hangzhou did not show significant differences among socioeconomic groups in terms of park access variations. Barriers to the development of the study of the dynamics of access inequity are mainly due to the availability of multi-temporal park distribution data and place-based socioeconomic data [6]. In the future, multidisciplinary cooperation with training and skills presents an opportunity to make full use of multi-source data, detect the reasons that may cause inequity, and to help evaluate current policies efficiency through dynamic analyses. City planners and policymakers should focus on those areas and groups of people experiencing decreasing accessibility and those priority zones that have already benefited from effectively implemented initiatives to ameliorate access inequity. We found very few studies in large areas of LMICs and the possible reasons that most of the empirical studies on UGS equity are located in western Europe, China, Australia, and the US might be local urbanization processes and the high demand of UGS for public health [58,60]. Since the US and China have been the major contributors to the publications in this field and have accumulated spatially explicit studies for individual cities, one of the pressing research areas is the comparison of park accessibility and the changes in accessibility among socioeconomic groups. Furthermore, the urban land structure could significantly influence the provision of UGS [82]. Thus, taking the different spatiotemporal patterns of urbanization in China and the US into account might enrich the research on spatial and temporal access inequity to UGS.

Advances in research on UGS access inequity have profoundly enhanced our understanding of the effects of the provision of UGS on social inequity. However, research on the contribution of interventions to improve access inequity remains a nascent field with substantial knowledge deficits. Hunter et al. [83] identified 6997 articles when reviewing the equity effect of urban green space interventions (only 38 articles were finally included) and found that there was too little evidence to draw firm conclusions. Both the political and scientific communities have acknowledged the importance of UGS, and it is the practical intervention strategies that could tackle the inequity issue in local urban contexts. Furthermore, to avoid exacerbated inequity, local governments must propose new initiatives, especially projects that would benefit regions with deficit residents. Cost-effective ways of green infrastructure network optimization, such as nature-based solutions, should be adopted to increase access to green spaces [84]. However, some cities may not prioritize urban planning in the creation of UGS and might prioritize industrial development and economic growth. The access equity issue may also be compromised by real-estate investment. Therefore, the governance, planning, and maintenance measures regarding the construction and redevelopment of UGS should not be detached from the urbanization traits. In particular, the temporal and spatial patterns of urban land expansion and urban population growth could help us to understand the formation mechanism of access inequity. Local interventions should be proposed based on public surveys and in situ research to guarantee the effectiveness of specific measures [85]. UGS access is not a single-dimension issue of green infrastructure, and therefore, the corresponding adjustment of a city's non-biotic or gray structures, especially the transport system, should be appropriately incorporated into the UGS planning system [60]. Sound UGS inequity mitigation policies, sensitive to the demographic distribution of income, age, ethnicity, or other social vulnerabilities, are much needed to effectively narrow the inequity gap. In the future, a combination of quantitative and qualitative methods might be essential to evaluate the effectiveness of policies and help policymakers find the pathways through which the inequity-intensifying impacts emerge. Scholarly local studies focusing on a given policy from a specific disciplinary view, and interdisciplinary considerations by the architectural, engineering, and construction professions are needed to enrich the research and develop new practical guidelines for urban green space planning. More studies are needed to uncover the access inequity issues for cities of LMICs with high populations concentrated in their urban areas.

5. Conclusions

We reported a systematic mapping of access inequity research based on a sample of 49 empirical studies screened from 563 selected papers. Although the scale of cities with UGS access inequity varies between countries, large cities (more than 1,000,000 population), especially in LMICs, are particularly affected. Disparities of UGS were found among different SES groups, and access inequity was investigated in different geographical settings. Across the reviewed papers, analyses on mitigating interventions are sparse. Using CA methodology, we found the critical trends, knowledge gaps, and clusters of this research field and underscored the incorporation of the UGS access inequity factor into urban planning and management to avoid exacerbated pressures on the sustainable development of cities. We provided the explanatory framework for the interaction between access inequity and local mitigating measures and have called for interdisciplinary cooperation by architectural, engineering, and construction professionals to reduce access inequity to UGS. In future research, it will be necessary to conduct comparative studies in various cities across different countries or continents for a comprehensive understanding of the factors that may influence the mitigation of this inequity.

Supplementary Materials: The following are available online at https://www.mdpi.com/article/10.3390/su14031062/s1. Figure S1: The time axis of quantitative models adopted for accessibility study and the number (in color) of publications for three typical aspects of access (quantity, proximity, and quality), Table S1: The inertia and contribution of individual dimensions, Table S2: The statistics of variables for the correspondence analysis in Figure 3, Table S3: Number of observational studies per different measure of access to urban green space, Table S4: Number of observational studies per different measure of SES, Table S5: Number of observational studies that included intervention analysis per different measure of exposure to urban green space, Table S6: Number of observational studies for each category of country level per different measure of urban green space access, Table S7: Number of observational studies for each category of temporal scale assessment per different measure of exposure to urban green space, Table S8: Number of observational studies for each category of spatial scale assessment per different measure of exposure to urban green space.

Author Contributions: Conceptualization, Y.S.; methodology, Y.S. and C.X.; validation, Y.S. and C.X.; formal analysis, Y.S.; investigation, Y.S. and X.K.; data curation, Y.S.; writing—original draft preparation, Y.S., S.S. and H.T.; writing—review and editing, Y.S., S.S., H.T. and C.X. All authors have read and agreed to the published version of the manuscript.

Funding: This study was supported by the National Natural Science Foundation of China (41801058) and the Fundamental Research Funds for the Central Universities (BLX201906). Somidh Saha thanks the German Aerospace Center (Das Deutsche Zentrum für Luft-und Raumfahrt e.V.) and The Federal Ministry of Education and Research of Germany (Das Bundesministerium für Bildung und Forschung—BMBF) for providing him a financial support (GrüneLunge project–funding reference number 01LR1726A, https://www.projekt-gruenelunge.de/ accessed on 25 November 2021).

Institutional Review Board Statement: Not applicable.

Informed Consent Statement: Not applicable.

Data Availability Statement: The data presented in this study are available from the corresponding author upon reasonable request.

Acknowledgments: We would like to thank anonymous reviewers for their constructive comments.

Conflicts of Interest: The authors declare no conflict of interest.

References

1. United Nations. *World Urbanization Prospects: The 2014 Revision*; United Nations Department of Economics and Social Affairs, Population Division: New York, NY, USA, 2015.
2. Demuzere, M.; Orru, K.; Heidrich, O.; Olazabal, E.; Geneletti, D.; Orru, H.; Bhave, A.G.; Mittal, N.; Feliu, E.; Faehnle, M. Mitigating and adapting to climate change: Multi-functional and multi-scale assessment of green urban infrastructure. *J. Environ. Manag.* **2014**, *146*, 107–115. [CrossRef] [PubMed]
3. Panno, A.; Carrus, G.; Lafortezza, R.; Mariani, L.; Sanesi, G. Nature-based solutions to promote human resilience and wellbeing in cities during increasingly hot summers. *Environ. Res.* **2017**, *159*, 249–256. [CrossRef]
4. Meerow, S.; Newell, J.P. Spatial planning for multifunctional green infrastructure: Growing resilience in Detroit. *Landsc. Urban Plan.* **2017**, *159*, 62–75. [CrossRef]
5. Tost, H.; Reichert, M.; Braun, U.; Reinhard, I.; Peters, R.; Lautenbach, S.; Hoell, A.; Schwarz, E.; Ebner-Priemer, U.; Zipf, A.; et al. Neural correlates of individual differences in affective benefit of real-life urban green space exposure. *Nat. Neurosci.* **2019**, *22*, 1389–1393. [CrossRef] [PubMed]
6. Collins, R.M.; Spake, R.; Brown, K.A.; Ogutu, B.O.; Smith, D.; Eigenbrod, F. A systematic map of research exploring the effect of greenspace on mental health. *Landsc. Urban Plan.* **2020**, *201*, 103823. [CrossRef]
7. Rigolon, A. A complex landscape of inequity in access to urban parks: A literature review. *Landsc. Urban Plan.* **2016**, *153*, 160–169. [CrossRef]
8. Song, Y.; Chen, B.; Kwan, M.-P. How does urban expansion impact people's exposure to green environments? A comparative study of 290 Chinese cities. *J. Clean. Prod.* **2020**, *246*, 119018. [CrossRef]
9. Kong, X.; Sun, Y.; Xu, C. Effects of Urbanization on the Dynamics and Equity of Access to Urban Parks from 2000 to 2015 in Beijing, China. *Forests* **2021**, *12*, 1796. [CrossRef]
10. Wu, L.; Rowe, P.G. Green space progress or paradox: Identifying green space associated gentrification in Beijing. *Landsc. Urban Plan.* **2022**, *219*, 104321. [CrossRef]
11. Wu, L.; Kim, S.K. Does socioeconomic development lead to more equal distribution of green space? Evidence from Chinese cities. *Sci. Total Environ.* **2021**, *757*, 143780. [CrossRef]
12. Rigolon, A.; Browning, M.; Jennings, V. Inequities in the quality of urban park systems: An environmental justice investigation of cities in the United States. *Landsc. Urban Plan.* **2018**, *178*, 156–169. [CrossRef]
13. Kabisch, N.; Haase, D. Green justice or just green? Provision of urban green spaces in Berlin, Germany. *Landsc. Urban Plan.* **2014**, *122*, 129–139. [CrossRef]
14. Xiao, Y.; Wang, Z.; Li, Z.G.; Tang, Z.L. An assessment of urban park access in Shanghai—Implications for the social equity in urban China. *Landsc. Urban Plan.* **2017**, *157*, 383–393. [CrossRef]
15. Dai, D. Racial/ethnic and socioeconomic disparities in urban green space accessibility: Where to intervene? *Landsc. Urban Plan.* **2011**, *102*, 234–244. [CrossRef]
16. Mitchell, R.; Popham, F. Effect of exposure to natural environment on health inequalities: An observational population study. *Lancet* **2008**, *372*, 1655–1660. [CrossRef]
17. McCormack, G.R.; Rock, M.; Toohey, A.M.; Hignell, D. Characteristics of urban parks associated with park use and physical activity: A review of qualitative research. *Health Place* **2010**, *16*, 712–726. [CrossRef]
18. Schipperijn, J.; Bentsen, P.; Troelsen, J.; Toftager, M.; Stigsdotter, U.K. Associations between physical activity and characteristics of urban green space. *Urban For. Urban Green.* **2013**, *12*, 109–116. [CrossRef]
19. Abercrombie, L.C.; Sallis, J.F.; Conway, T.L.; Frank, L.D.; Saelens, B.E.; Chapman, J.E. Income and racial disparities in access to public parks and private recreation facilities. *Am. J. Pre. Med.* **2008**, *34*, 9–15. [CrossRef]
20. Almohamad, H.; Knaack, A.L.; Habib, B.M. Assessing Spatial Equity and Accessibility of Public Green Spaces in Aleppo City, Syria. *Forests* **2018**, *9*, 706. [CrossRef]
21. Arroyo-Johnson, C.; Woodward, K.; Milam, L.; Ackermann, N.; Komaie, G.; Goodman, M.S.; Hipp, J.A. Still Separate, Still Unequal: Social Determinants of Playground Safety and Proximity Disparities in St. Louis. *J. Urban Health* **2016**, *93*, 627–638. [CrossRef] [PubMed]
22. Astell-Burt, T.; Feng, X.Q.; Mavoa, S.; Badland, H.M.; Giles-Corti, B. Do low-income neighbourhoods have the least green space? A cross-sectional study of Australia's most populous cities. *BMC Public Health* **2014**, *14*, 292. [CrossRef] [PubMed]
23. Bahrini, F.; Bell, S.; Mokhtarzadeh, S. The relationship between the distribution and use patterns of parks and their spatial accessibility at the city level: A case study from Tehran, Iran. *Urban For. Urban Green.* **2017**, *27*, 332–342. [CrossRef]
24. Barbosa, O.; Tratalos, J.A.; Armsworth, P.R.; Davies, R.G.; Fuller, R.A.; Johnson, P.; Gaston, K.J. Who benefits from access to green space? A case study from Sheffield, UK. *Landsc. Urban Plan.* **2007**, *83*, 187–195. [CrossRef]
25. Bruton, C.M.; Floyd, M.F. Disparities in Built and Natural Features of Urban Parks: Comparisons by Neighborhood Level Race/Ethnicity and Income. *J. Urban Health* **2014**, *91*, 894–907. [CrossRef] [PubMed]

26. Chen, S.L.; Christensen, K.M.; Li, S.J. A comparison of park access with park need for children: A case study in Cache County, Utah. *Landsc. Urban Plan.* **2019**, *187*, 119–128. [CrossRef]
27. Comber, A.; Brunsdon, C.; Green, E. Using a GIS-based network analysis to determine urban greenspace accessibility for different ethnic and religious groups. *Landsc. Urban Plan.* **2008**, *86*, 103–114. [CrossRef]
28. Cradock, A.L.; Kawachi, I.; Colditz, G.A.; Hannon, C.; Melly, S.J.; Wiecha, J.L.; Gortmaker, S.L. Playground safety and access in Boston neighborhoods. *Am. J. Prev. Med* **2005**, *28*, 357–363. [CrossRef] [PubMed]
29. Dadvand, P.; de Nazelle, A.; Figueras, F.; Basagana, X.; Su, J.; Amoly, E.; Jerrett, M.; Vrijheid, M.; Sunyer, J.; Nieuwenhuijsen, M.J. Green space, health inequality and pregnancy. *Environ. Int.* **2012**, *40*, 110–115. [CrossRef]
30. de Mola, U.L.; Ladd, B.; Duarte, S.; Borchard, N.; La Rosa, R.A.; Zutta, B. On the Use of Hedonic Price Indices to Understand Ecosystem Service Provision from Urban Green Space in Five Latin American Megacities. *Forests* **2017**, *8*, 478. [CrossRef]
31. Engelberg, J.K.; Conway, T.L.; Geremia, C.; Cain, K.L.; Saelens, B.E.; Glanz, K.; Frank, L.D.; Sallis, J.F. Socioeconomic and race/ethnic disparities in observed park quality. *BMC Public Health* **2016**, *16*, 395. [CrossRef]
32. Feng, S.; Chen, L.D.; Sun, R.H.; Feng, Z.Q.; Li, J.R.; Khan, M.S.; Jing, Y.C. The Distribution and Accessibility of Urban Parks in Beijing, China: Implications of Social Equity. *Int. J. Environ. Res. Public Health* **2019**, *16*, 4894. [CrossRef] [PubMed]
33. Gu, X.K.; Tao, S.Y.; Dai, B. Spatial accessibility of country parks in Shanghai, China. *Urban For. Urban Green.* **2017**, *27*, 373–382. [CrossRef]
34. Guo, S.H.; Song, C.; Pei, T.; Liu, Y.X.; Ma, T.; Du, Y.Y.; Chen, J.; Fan, Z.D.; Tang, X.L.; Peng, Y.; et al. Accessibility to urban parks for elderly residents: Perspectives from mobile phone data. *Landsc. Urban Plan.* **2019**, *191*, 103642. [CrossRef]
35. He, S.W.; Wu, Y.L.; Wang, L. Characterizing Horizontal and Vertical Perspectives of Spatial Equity for Various Urban Green Spaces: A Case Study of Wuhan, China. *Front. Public Health* **2020**, *8*, 10. [CrossRef]
36. Hoffimann, E.; Barros, H.; Ribeiro, A.I. Socioeconomic Inequalities in Green Space Quality and Accessibility-Evidence from a Southern European City. *Int. J. Environ. Res. Public Health* **2017**, *14*, 916. [CrossRef]
37. Iraegui, E.; Augusto, G.; Cabral, P. Assessing Equity in the Accessibility to Urban Green Spaces According to Different Functional Levels. *ISPRS Int. Geo-Inf.* **2020**, *9*, 308. [CrossRef]
38. Jenkins, G.R.; Yuen, H.K.; Rose, E.J.; Maher, A.I.; Gregory, K.C.; Cotton, M.E. Disparities in Quality of Park Play Spaces between Two Cities with Diverse Income and Race/Ethnicity Composition: A Pilot Study. *Int. J. Environ. Res. Public Health* **2015**, *12*, 8009–8022. [CrossRef] [PubMed]
39. Kamel, A.A.; Ford, P.B.; Kaczynski, A.T. Disparities in park availability, features, and characteristics by social determinants of health within a US-Mexico border urban area. *Prev. Med.* **2014**, *69*, S111–S113. [CrossRef] [PubMed]
40. Knapp, M.; Gustat, J.; Darensbourg, R.; Myers, L.; Johnson, C. The Relationships between Park Quality, Park Usage, and Levels of Physical Activity in Low-Income, African American Neighborhoods. *Int. J. Environ. Res. Public Health* **2019**, *16*, 85. [CrossRef] [PubMed]
41. La Rosa, D.; Takatori, C.; Shimizu, H.; Privitera, R. A planning framework to evaluate demands and preferences by different social groups for accessibility to urban greenspaces. *Sust. Cities Soc.* **2018**, *36*, 346–362. [CrossRef]
42. Lara-Valencia, F.; Garcia-Perez, H. Space for equity: Socioeconomic variations in the provision of public parks in Hermosillo, Mexico. *Local Environ.* **2015**, *20*, 350–368. [CrossRef]
43. Lara-Valencia, F.; Garcia-Perez, H. Disparities in the provision of public parks in neighbourhoods with varied Latino composition in the Phoenix Metropolitan Area. *Local Environ.* **2018**, *23*, 1107–1120. [CrossRef]
44. Lin, B.; Meyers, J.; Barnett, G. Understanding the potential loss and inequities of green space distribution with urban densification. *Urban For. Urban Green.* **2015**, *14*, 952–958. [CrossRef]
45. Manta, S.W.; Reis, R.S.; Benedetti, T.R.B.; Rech, C.R. Public open spaces and physical activity: Disparities of resources in Florianopolis. *Rev. Saude Publica* **2019**, *53*, 112. [CrossRef]
46. Nero, B.F. Urban green space dynamics and socio-environmental inequity: Multi-resolution and spatiotemporal data analysis of Kumasi, Ghana. *Int. J. Remote Sens.* **2017**, *38*, 6993–7020. [CrossRef]
47. Park, J.H.; Lee, D.K.; Park, C.; Kim, H.G.; Jung, T.Y.; Kim, S. Park Accessibility Impacts Housing Prices in Seoul. *Sustainability* **2017**, *9*, 185. [CrossRef]
48. Rahman, K.M.A.; Zhang, D.F. Analyzing the Level of Accessibility of Public Urban Green Spaces to Different Socially Vulnerable Groups of People. *Sustainability* **2018**, *10*, 3917. [CrossRef]
49. Reyes, M.; Paez, A.; Morency, C. Walking accessibility to urban parks by children: A case study of Montreal. *Landsc. Urban Plan.* **2014**, *125*, 38–47. [CrossRef]
50. Sathyakumar, V.; Ramsankaran, R.; Bardhan, R. Linking remotely sensed Urban Green Space (UGS) distribution patterns and Socio-Economic Status (SES)-A multi-scale probabilistic analysis based in Mumbai, India. *Gisci Remote Sens.* **2019**, *56*, 645–669. [CrossRef]
51. Schule, S.A.; Gabriel, K.M.A.; Bolte, G. Relationship between neighbourhood socioeconomic position and neighbourhood public green space availability: An environmental inequality analysis in a large German city applying generalized linear models. *Int. J. Hyg. Environ. Health* **2017**, *220*, 711–718. [CrossRef]
52. Shen, Y.A.; Sun, F.; Che, Y.Y. Public green spaces and human wellbeing: Mapping the spatial inequity and mismatching status of public green space in the Central City of Shanghai. *Urban For. Urban Green.* **2017**, *27*, 59–68. [CrossRef]

53. Sugiyama, T.; Villanueva, K.; Knuiman, M.; Francis, J.; Foster, S.; Wood, L.; Giles-Corti, B. Can neighborhood green space mitigate health inequalities? A study of socio-economic status and mental health. *Health Place* **2016**, *38*, 16–21. [CrossRef] [PubMed]
54. Tan, C.D.; Tang, Y.H.; Wu, X.F. Evaluation of the Equity of Urban Park Green Space Based on Population Data Spatialization: A Case Study of a Central Area of Wuhan, China. *Sensors* **2019**, *19*, 2929. [CrossRef]
55. Tan, P.Y.; Samsudin, R. Effects of spatial scale on assessment of spatial equity of urban park provision. *Landsc. Urban Plan.* **2017**, *158*, 139–154. [CrossRef]
56. Tian, Y.H.; Jim, C.Y.; Liu, Y.Q. Using a Spatial Interaction Model to Assess the Accessibility of District Parks in Hong Kong. *Sustainability* **2017**, *9*, 1924. [CrossRef]
57. Tu, X.Y.; Huang, G.L.; Wu, J.G. Contrary to Common Observations in the West, Urban Park Access Is Only Weakly Related to Neighborhood Socioeconomic Conditions in Beijing, China. *Sustainability* **2018**, *10*, 1115. [CrossRef]
58. Wei, F. Greener urbanization? Changing accessibility to parks in China. *Landsc. Urban Plan.* **2017**, *157*, 542–552. [CrossRef]
59. Weiss, C.C.; Purciel, M.; Bader, M.; Quinn, J.W.; Lovasi, G.; Neckerman, K.M.; Rundle, A.G. Reconsidering Access: Park Facilities and Neighborhood Disamenities in New York City. *J. Urban Health* **2011**, *88*, 297–310. [CrossRef]
60. Wende, H.E.W.; Zarger, R.K.; Mihelcic, J.R. Accessibility and usability: Green space preferences, perceptions, and barriers in a rapidly urbanizing city in Latin America. *Landsc. Urban Plan.* **2012**, *107*, 272–282. [CrossRef]
61. Xu, C.; Haase, D.; Pribadi, D.O.; Pauleit, S. Spatial variation of green space equity and its relation with urban dynamics: A case study in the region of Munich. *Ecol. Indic.* **2018**, *93*, 512–523. [CrossRef]
62. Yang, J.; Li, C.; Li, Y.C.; Xi, J.H.; Ge, Q.S.; Li, X.M. Urban green space, uneven development and accessibility: A case of Dalian's Xigang District. *Chin. Geogr. Sci.* **2015**, *25*, 644–656. [CrossRef]
63. Zhang, J.G.; Cheng, Y.Y.; Wei, W.; Zhao, B. Evaluating Spatial Disparity of Access to Public Parks in Gated and Open Communities with an Improved G2SFCA Model. *Sustainability* **2019**, *11*, 5910. [CrossRef]
64. Zhou, X.L.; Kim, J. Social disparities in tree canopy and park accessibility: A case study of six cities in Illinois using GIS and remote sensing. *Urban For. Urban Green.* **2013**, *12*, 88–97. [CrossRef]
65. Brown, G.; Brabyn, L. An analysis of the relationships between multiple values and physical landscapes at a regional scale using public participation GIS and landscape character classification. *Landsc. Urban Plan.* **2012**, *107*, 317–331. [CrossRef]
66. Dembek, K.; Sivasubramaniam, N.; Chmielewski, D.A. A Systematic Review of the Bottom/Base of the Pyramid Literature: Cumulative Evidence and Future Directions. *J. Bus. Ethics* **2019**, *165*, 365–382. [CrossRef]
67. Caula, S.; Marty, P.; Martin, J.-L. Seasonal variation in species composition of an urban bird community in Mediterranean France. *Landsc. Urban Plan.* **2008**, *87*, 1–9. [CrossRef]
68. Zabret, K.; Šraj, M. Evaluating the Influence of Rain Event Characteristics on Rainfall Interception by Urban Trees Using Multiple Correspondence Analysis. *Water* **2019**, *11*, 2659. [CrossRef]
69. Greenacre, M. Correspondence analysis of raw data. *Ecology* **2010**, *91*, 958–963. [CrossRef]
70. James, K.L.; Randall, N.P.; Haddaway, N.R. A methodology for systematic mapping in environmental sciences. *Environ. Evid.* **2016**, *5*, 7. [CrossRef]
71. Lê, S.; Josse, J.; Husson, F. FactoMineR: An R package for multivariate analysis. *J. Stat. Softw.* **2008**, *25*, 1–18. [CrossRef]
72. Ibes, D.C. A multi-dimensional classification and equity analysis of an urban park system: A novel methodology and case study application. *Landsc. Urban Plan.* **2015**, *137*, 122–137. [CrossRef]
73. Xing, L.; Liu, Y.; Liu, X.; Wei, X.; Mao, Y. Spatio-temporal disparity between demand and supply of park green space service in urban area of Wuhan from 2000 to 2014. *Habitat Int.* **2018**, *71*, 49–59. [CrossRef]
74. Martinez-Vazquez, J.; Qiao, B.; Zhang, L. The role of provincial policies in fiscal equalization outcomes in China. *China Rev.* **2008**, *8*, 135–167.
75. Radke, J.; Mu, L. Spatial Decompositions, Modeling and Mapping Service Regions to Predict Access to Social Programs. *Geogr. Inf. Sci.* **2000**, *6*, 105–112. [CrossRef]
76. Rigolon, A.; Flohr, T. Access to Parks for Youth as an Environmental Justice Issue: Access Inequalities and Possible Solutions. *Buildings* **2014**, *4*, 69–94. [CrossRef]
77. Lee, G.; Hong, I. Measuring spatial accessibility in the context of spatial disparity between demand and supply of urban park service. *Landsc. Urban Plan.* **2013**, *119*, 85–90. [CrossRef]
78. Li, L.; Li, J.; Jiang, Z.; Zhao, L.; Zhao, P. Methods of Population Spatialization Based on the Classification Information of Buildings from China's First National Geoinformation Survey in Urban Area: A Case Study of Wuchang District, Wuhan City, China. *Sensors* **2018**, *18*, 2558. [CrossRef] [PubMed]
79. Jim, C.Y.; Chen, W.Y. Recreation–amenity use and contingent valuation of urban greenspaces in Guangzhou, China. *Landsc. Urban Plan.* **2006**, *75*, 81–96. [CrossRef]
80. Raciti, S.M.; Hutyra, L.R.; Newell, J.D. Mapping carbon storage in urban trees with multi-source remote sensing data: Relationships between biomass, land use, and demographics in Boston neighborhoods. *Sci. Total Environ.* **2014**, *500–501*, 72–83. [CrossRef]
81. Mitchell, M.G.E.; Johansen, K.; Maron, M.; McAlpine, C.A.; Wu, D.; Rhodes, J.R. Identification of fine scale and landscape scale drivers of urban aboveground carbon stocks using high-resolution modeling and mapping. *Sci. Total Environ.* **2018**, *622–623*, 57–70. [CrossRef]

82. Boulton, C.; Dedekorkut-Howes, A.; Byrne, J. Factors shaping urban greenspace provision: A systematic review of the literature. Landsc. *Urban Plan.* **2018**, *178*, 82–101. [CrossRef]
83. Hunter, R.F.; Cleland, C.; Cleary, A.; Droomers, M.; Wheeler, B.W.; Sinnett, D.; Nieuwenhuijsen, M.J.; Braubach, M. Environmental, health, wellbeing, social and equity effects of urban green space interventions: A meta-narrative evidence synthesis. *Environ. Int.* **2019**, *130*, 104923. [CrossRef] [PubMed]
84. Faivre, N.; Fritz, M.; Freitas, T.; de Boissezon, B.; Vandewoestijne, S. Nature-Based Solutions in the EU: Innovating with nature to address social, economic and environmental challenges. *Environ. Res.* **2017**, *159*, 509–518. [CrossRef] [PubMed]
85. Stessens, P.; Canters, F.; Huysmans, M.; Khan, A.Z. Urban green space qualities: An integrated approach towards GIS-based assessment reflecting user perception. *Land Use Policy* **2020**, *91*, 104319. [CrossRef]

 *sustainability*

Article

Predicting Detached Housing Vacancy: A Multilevel Analysis

Jaekyung Lee [1], Galen Newman [2],* and Changyeon Lee [3]

[1] Department of Urban Design and Planning, Hongik University, Seoul 04066, Korea; jklee1@hongik.ac.kr
[2] Department of Landscape Architecture and Urban Planning, Texas A&M University, College Station, TX 77843, USA
[3] Department of Urban Engineering, Jeonbuk National University, Jeonju 54896, Korea; idealcity@jbnu.ac.kr
* Correspondence: gnewman@tamu.edu; Tel.: +82-979-862-4320

Abstract: Urban shrinkage is a critical issue in local small- and medium-sized cities in Korea. While there have been several studies to analyze the causes and consequences of vacancy increases, most have only focused on socioeconomic associations at larger scale and failed to consider individual housing level characteristics, primarily due to a lack of appropriate data. Based on data including 52,400 individual parcels, this study analyzes the primary contributors to vacant properties and their spatial distribution through a multilevel model design based on data for each parcel. Then, we identify areas at high risk of vacancy in the future to provide evidence to establish policies for improving the local environment. Results indicate that construction year, building structure, and road access conditions have a significant effect on vacant properties at the individual parcel level, and the presence of schools and hypermarket within 500 m are found to decrease vacant properties. Further, prediction outcomes show that the aged city center and areas with strict regulations on land use are expected to have a higher vacancy rate. These findings are used to provide a set of data-based revitalization strategies through the development of a vacancy prediction model.

Keywords: urban shrinkage; vacancy parcel data; multilevel analysis; predicting vacancy

Citation: Lee, J.; Newman, G.; Lee, C. Predicting Detached Housing Vacancy: A Multilevel Analysis. *Sustainability* 2022, 14, 922. https://doi.org/10.3390/su14020922

Academic Editors: Miguel Amado and Pierfrancesco De Paola

Received: 4 November 2021
Accepted: 12 January 2022
Published: 14 January 2022

Publisher's Note: MDPI stays neutral with regard to jurisdictional claims in published maps and institutional affiliations.

1. Introduction

Recent rapid urban growth has forced Korean cities to focus on physical and spatial expansion [1]. As such, planning policies related to growth management or smart growth are being developed with sustainability and development in mind [2]. Accordingly, a large number of development projects have been put forth in response to the growing demand for housing in the Seoul Metropolitan Area (SMA) of South Korea, such as the ambitious 3rd Generation New Towns project. In contrast, many small- and medium-sized cities and rural areas outside of the SMA are experiencing depopulation that first began in the 1980s. Of 230 Korean cities, 139 (60.4%) experienced depopulation between 1990 and 2015. While only 18.6% (20) of cities in the SMA lost population during this period, non-metropolitan areas also showed a 30.1% (119) decline, indicating urban shrinkage in these regions [3]. In particular, this trend is expected to worsen for small- and medium-sized cities due to continued low fertility rates (in fact, Korea has the lowest among OECD member countries) and aging of the population [4].

Urban decline can be a result of shrinkage and cause not only a long-term economic recession, but also socioeconomic problems including deteriorating neighborhoods and living environments, poor regional competitiveness, and high crime rates [5]. Another related issue is that of excessive amounts of vacant land—a physical manifestation of economic decline brough about by depopulation [6,7]. According to the 2017 Housing Census by Statistics Korea [8], vacant properties accounted for 7.4% (1,270,000) of the housing properties in the country at the time—a 3.4-fold increase in just 22 years from 360,000 (3.9%) in 1995. Statistics Korea forecasts that vacant properties will account for 10% (3 million) of housing properties by 2050.

If left unattended long-term, vacant properties can cause complex physical, social, and economic problems such as deteriorating residential conditions, poor esthetics, increasing crime and disasters, falling housing prices, and higher administrative costs. We must establish a framework to both upgrade and reuse vacant properties, especially given the projected rise in vacant properties. Housing demand will decrease beyond absolute housing shortages based on the current housing supply rate above 100% (total number of houses to total number of households ratio). This will be coupled with a decreasing population growth rate and a declining working-age population.

We understand that all cities seek to promote economic growth through the use of diverse resources. Among these resources, land is crucial—especially vacant land [9]. As noted, shrinking cities have a much larger influence on the socioeconomic fabric of affected societies. Among these problems, the issue of vacant properties is especially critical to address [10]. If we recognize vacant properties as both the *cause* and *result* of a declining city, and accordingly identify and manage this cause-and-effect relationship, it may be possible to preemptively prevent the vicious cycle of vacant properties and deteriorating cities. Under the hypothesis that a property becomes vacant/abandoned not only due to individual building characteristics but also regional economic, social, and environmental characteristics, the multilevel logistic regression model was used to extract the determinants of vacancies at different levels and predict the probability of vacancy in each building in the future.

The literature has proposed various definitions of "vacant properties" and their contributing factors across different scales—from city-based macroscopic to neighborhood- and individual housing unit-based local scales. Vacant land is said to include not only underperforming or unoccupied properties, but also neglected or abandoned commercial/industrial buildings that may pose a threat to public safety [10,11]. Vacant land may also include open spaces such as parks, farm sites, and properties with natural resource value [6].

Vacant housing in residential areas are especially worrisome. Although there is no agreed-upon concept and definition of vacant properties in Korea in the literature, the Act on Special Cases Concerning Unoccupied House or Small-Scale Housing Improvement, 2017, defines vacant properties as "a house not occupied or used for at least one year as of the date the Special Self-Governing City Mayor has verified that the house was occupied or used." Surveys into such properties in different regions have since been conducted routinely.

On the contributing factors of vacant properties, Scafidi et al. (1998) and Hillier et al. (2003) [12,13] found that housing characteristics and the socioeconomic characteristics of owners affected housing abandonment in declining cities across the US. They confirmed that, if housing properties have poor functional and economic conditions, or homeowners have poor financial conditions, they are more likely to be neglected and therefore become vacant. Bassett et al. (2006), Silverman et al. (2013), and Morckel et al. (2013) [14–16] found that the proportion of African-Americans, proportion of poorly educated, unemployment rates, poverty rates, age of the building, and housing prices are all related to housing abandonment; that is, a local area's physical, social, and economic characteristics may trigger housing abandonment. Morckel (2013) [16] conducted a multilayer analysis of Youngstown and Columbus to analyze the effect of the neighborhood scale on housing abandonment, and identified the relationship between vacant properties and three conditions in the target and nearby areas: market conditions, gentrification, and physical abandonment.

Further, local market and economic conditions may also affect the emergence of vacant properties. According to the real estate market, the main cause of vacant housing is a low demand for such properties. Such properties arise because of a mismatch between supply and demand under conditions of housing oversupply [17,18]. If the demand is lower than the housing supply, consumers improve the quality of their houses through housing filtering; in this process, houses with low competitiveness are abandoned [19,20]. If many new housing properties are supplied without changes in local housing demand, properties with low market values, such as old and poor-quality ones, will be left vacant and abandoned. This phenomenon evolves beyond a single area. If a specific area has a low level of market competitiveness, residents who can afford to move will do so, thus producing more vacant properties.

Studies identifying the spatial distribution of vacant properties across the city and predicting areas at higher risk of vacancy are rare; a deep-learning-based vacancy prediction model was only developed in 2016 [10,13,21]. Hillier et al. [13] identified the determinants of vacancy and predicted housing abandonment in Philadelphia using a logistic regression model. They found that identifying the principal predictor of abandonment is essential for establishing long-term strategies to protect housing stock. Furthermore, they noted that the causes of vacancies differ depending on the scale.

Newman et al. (2016) and Lee and Newman (2017) [10,21] developed a neural-network-based vacancy prediction model. By comparing four different model accuracy tests, they proved that the model can be a suitable alternative for predicting future possible urban vacancy dynamics and conditions. By quantifying and comparing factors that occur in a growing city (Fort Wort, TX, USA) and a declining city (Chicago, IL, USA), they found that while transportation-related variables in the declining city are major contributors to vacancy, socioeconomic variables such as unemployment rate, housing price, and race have stronger influence in a growing city. With the latest technological advances that can collect and analyze various types of data, scholars are beginning to analyze and predict vacancy patterns based on the deep-learning-based prediction model. However, most deep-learning-based predictive models require time-series data, and it is difficult to conduct a more detailed analysis because the analysis must be conducted with raster-based pixel units. Since this study obtained 2020 parcel-level vacancy data, multilevel logistic regression model was utilized to predict the vacancy possibility of each parcel.

The primary objectives of this research are twofold: (1) to accurately analyze contributors to vacant properties and spatial distribution through a multilevel model design based on data for individual parcels, and (2) to provide evidence to establish policies to boost the local area by predicting areas at risk of vacancy. Our study differs from the prevailing works on vacant properties in three ways. First, prior studies are limited in their analysis of the characteristics of vacant properties in a more specific unit of land because they rely on data at the administrative district scale, given the challenges in obtaining proper data. For example, Morckel (2013) [16] used tax-delinquent information on individual properties as a proxy for abandonment, while Zou and Wang (2020) [22] detected vacant houses using high-resolution remote sensing images. We conducted a more detailed and accurate spatial analysis based on data from approximately 32,000 individual parcels in Jeonju, a small- and medium-sized local city in western South Korea experiencing a rapid decline.

Second, while some scholars use data from individual buildings, most limit their analysis to only a few areas in the city or focus on physical characteristics such as construction year and architectural type. We included not only the physical characteristics of individual parcels, but also the characteristics of neighborhood residential conditions such as access to amenities as key explanatory variables, not limiting ourselves to a certain small area within the city. Williams et al. (2019) [23] highlighted the importance of access to amenities such as schools, grocery stores, medical facilities, and malls to allow communities from ghost cities (i.e., cities with housing vacancies) to flourish. Access to public transport also increases housing value [24–26]. Accordingly, variables such as access to a variety of amenities, such as elementary schools, hypermarkets, and hospitals, the presence of unwanted facilities such as waste treatment plants, and vacant properties were employed in our analysis.

Third, vacant properties may be related to conditions differently across different scales. While Accordino and Johnson (2000) [20] showed that specific housing conditions and neighboring conditions influence housing properties, most other scholars seem to focus on the socioeconomic status of the entire city level due to the lack of vacancy-related data. This includes population and employment status rather than neighborhood conditions such as service accessibilities, and the physical conditions of each property, such as structures and the year built.

By investigating each house's conditions, we included not only the local area's socioeconomic characteristics but also the unique characteristics of individual buildings and access to infrastructure related to neighborhood environments as key explanatory variables. We employed a multilevel logistic regression model to identify parcels at high risk of vacancy in the future, instead of using past data to analyze trends and suggest policy alternatives. In addition to identifying parcels at high risk of vacancy, we identify high risk of vacancy clusters at the regional scale within the study area by examining their spatial distribution with hotspot analysis. Within this process, we present a reference to establish more realistic and future-oriented policies.

While there has been a concerted effort to study and understand how vacant properties can be used, without identifying the causes and consequences of the increase in vacancy, it is difficult to systematically manage and use these properties. In the context of our study—South Korea—we note that while the Korean government creates the available data on vacant housing property through the Housing Census of Statistics Korea, the information is only a simple aggregate that includes spatial ranges of cities or census tract levels. These data lack accuracy and specificity in identifying the *number* and *status* of vacant properties, thus rendering them ineffective for analyzing the microscopic and accurate causes of vacant properties that take regional heterogeneity into account. In the literature, for example, while Jeon as well as Kim and Park [27,28] conducted their studies at the scale of individual buildings, they failed to identify the spatial distribution or cause of vacant properties across the city; further, they did not consider access to amenities and the presence of nearby vacant properties.

2. Materials and Methods

2.1. Study Area

To examine urban vacancy patterns and their causes, we examined the case of Jeonju, located in the North Jeolla Province, South Korea (Figure 1). The extinction risk index was considered to measure the urban shrinkage. The index was first introduced to measure the urban shrinkage of Japan [29]. It is the value dividing the female population aged 20 to 39 years in a region by the population aged 65 years and above in the region. According to an analysis of trends in the local extinction risk index for 261 local governments across Korea, Jeonju was rated 0.67 as of 2020, which is quite low. However, as its risk has increased sharply since 2000, urgent measures need to be taken (see Table 1 for the extinction risk index). Lee [30,31] analyzed Korea's local extinction and found that if the value is under 0.5, depopulation will continue to stagnate for over 20 years.

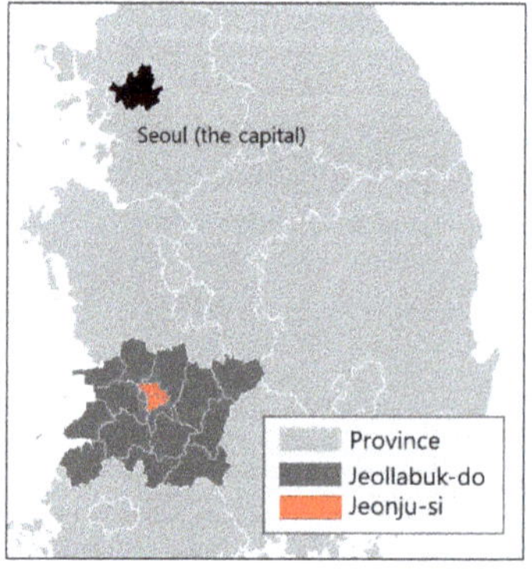

Total Population	657,661
0–14yrs	13.4 (%)
15–64yrs	71.6 (%)
65yrs–	15.0 (%)
Area	206.2 (km²)
Total household	285,227
Employment Rate	58.3%

Figure 1. Location and demographics of the study area.

Table 1. Location extinction risk index in the study area.

	2000	**2010**	**2020**	**2010–2020 Rate of Change**	**Local Extinction Risk Index**
Jeonju	2.5	1.3	0.7	73.4%	Low extinction risk: ≥1.5
North Jeolla Province	1.3	0.7	0.5	63.6%	Moderate extinction risk: 1.0–1.5 High extinction risk: 0.5–1.0
Nationwide	2.1	1.1	0.6	71.5%	Very high extinction risk: <0.5

As vacant housing properties caused by population outflows are becoming a serious social issue, Jeonju is creating strategies to reuse vacant properties and explore how to manage and support these properties by reviewing relevant laws and ordinances. In 2020, Jeonju collaborated with the Land and Geospatial Informatrix Corporation to establish a database on vacant properties. They conducted a survey of vacant properties left unattended for over one year in the city and began to seek strategies to use these properties, taking into account aging and use of unoccupied properties.

We analyzed the factors contributing to vacant properties in 31,810 detached housing parcels in Jeonju. We then predicted the areas likely to have vacant properties in the future using a multilevel logistic regression model and explored measures to tackle housing vacancy issues. Our analysis provides implications for a more objective and systematic response, as we examined the status of vacant properties with two different levels of variables: (1) individual building level such as parcel size, shape, construction year, and access to infrastructure, and (2) the Dong, which is the smallest administrative unit in Korea, sharing relatively homogeneous socioeconomic conditions such as population density, employed population, and fertility rate [32].

2.2. Housing Vacancy Database

According to the "Act on Special Cases Concerning Unoccupied Houses or Small-scale Housing Improvement", a vacant house is defined as a property that has been unoccupied for more than one year. Since it is difficult to keep track of properties that have been vacant for over one year, we identified the occupation status and building conditions of current housing. To obtain the vacancy data, the most accurate way to build data for vacant properties is to investigate the status of all buildings in the study area and determine whether they are vacant and how old they are. However, we surveyed the status of vacant properties to build a more effective data-limited budget using the following three steps (see Figure 2):

(1) Extract buildings expected to be vacant properties based on data on water and electricity consumption in individual residential buildings;
(2) Visit and investigate the buildings extracted from (1), and confirm whether they are actually vacant properties;
(3) Identify the buildings' conditions such as abandonment, aging, owner, access, and the possibility of potential future reuse for buildings confirmed to be vacant properties.

Figure 2. Vacant property's survey process.

Based on the parcel numbering unit code for vacant properties derived from the above process, we established a building database along with building data from the geographic information system. The established building database includes various characteristics, including building identification number, parcel number, building structure, land area, gross floor area, violated building, building area, height, and construction year [1]. These data can be used to analyze contributors to vacant properties. From this process, a total of 52,429 parcels were extracted and spatialized out of 88,404 in Jeonju, excluding those that could not be analyzed (including illegal ones) as their characteristic data were not included in the building register. Since some multifamily properties share water and electricity usage, the survey has a limitation to determine whether a house is vacant within multifamily housing. Thus, 31,810 detached house parcels were re-extracted and used for analysis, and 4.2% of them (1348) were confirmed to be vacant (Table 2).

Table 2. Total and analyzed single family parcels in numbers and ratio.

Total Parcels (%)	Total Vacant Properties (%)	Total Analyzed Parcels (%)	Vacant Properties in Analyzed Parcels (%)
52,429 (100%)	2046 (3.9%)	31,810 (60.7%)	1348 (4.2%)

As shown in Figure 3, vacant properties are densely concentrated around the historic city center. For the historic city center, large-scale housing development projects on the edges of the city in the late 1970s led to a decline in population in residential areas in Jeonju; the North Jeolla Provincial Office was relocated to the outskirts of the city in 2005, bringing the city's residential base into a further decline and exacerbating the issue of vacant properties. This reorganization of urban functions not only relocated or dispersed urban functions but forced residents to flow out to where the functions were relocated, thus increasing loss of employment and urban decline.

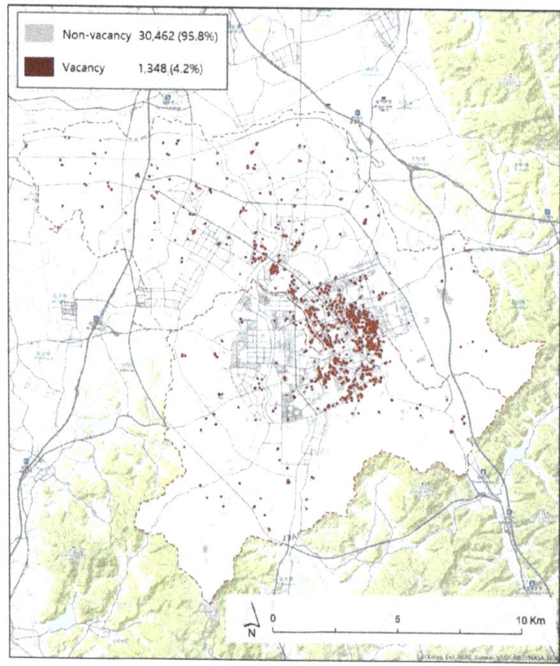

Figure 3. Distribution of parcel of detached house land and vacant properties in the study area.

Because we analyzed vacancies within individual parcels, vacancies were expressed in binary code as 1 and non-vacancies as 0. Since the dependent variable was binary, a binary logistic regression analysis was performed.

Based on vacant property data from the above process, we identified the physical, spatial, and demographic characteristics of vacant properties using a dataset from two different scales: (1) individual building level and (2) the Dong level. Based on these characteristics, we predict the probability of non-vacant buildings becoming vacant properties in the future, identify areas at risk of vacancy, and establish a framework to devise relevant policies in the future.

2.3. Variable Data

Considering the factors reviewed from previous studies, seventeen variables at two levels were used as factors contributing to vacancy: those representing local characteristics and those reflecting the characteristics of individual parcels (Table 3). Variables representing local characteristics included five variables, such as the percentage of secondary industry workers, population density, the percentage of old buildings and parks and green area per person. Variables reflecting the characteristics of individual parcels included twelve variables, such as construction year, building structure, type of parcel, and road access conditions [33]. Within individual parcel variables, the total number of parcels within a radius of 500 m from amenities, such as elementary schools and hypermarkets, and the total number of parcels within a radius of 500 m from vacant ones were identified by spatial analysis in order to consider the effect of neighborhood environments around parcels on vacant properties.

Among individual parcel data, there were 11 dummy variables (0 or 1) excluding parcel area and construction year, and the reinforced concrete structure was set as default and compared with block, wood, steel, and other structures in order to obtain a relative vacancy probability. Regarding access to key infrastructure related to neighborhood environments, the area influenced by amenities was set at a radius of 500 m based on the literature examining a walkable and service-accessible distance [34–36], and we identified whether a parcel was included within a radius of 500 m from the center of amenities. Parcels within a radius of 500 m from unwanted facilities were also included in this study's analysis because unwanted facilities such as waste treatment plants and wastewater discharge facilities may contribute to vacancies [37–39]. Over 50% of detached houses were located within a radius of 500 m from infrastructure for amenities such as elementary schools and hypermarkets.

Before analysis, a comparison of the characteristics of 31,810 detached house parcels confirmed that the percentage of wooden structures was 39.2% among vacant properties— 2.7 times higher than that of wooden structures among non-vacant properties. Road access conditions and age (construction year) had poorer vacant properties than other detached house parcels. In contrast, the type of parcel (regular/irregular) in vacant properties was found to be regular, which is expected to have a positive effect on the use of the properties in the future (Table 4).

Table 3. Characteristics of variables.

	Variable	Description	Source (Year)
Dependent variable	Vacancy		• Land and Geospatial Informatrix Corporation vacancy database (2020) • Seumteo Building Register (2020)
Dong level	• Percentage of secondary industry workers • Fertility rate • Population density • Percentage of old buildings • Park and green area per person	• Percentage of secondary industry workers (mining, manufacturing, and construction) among all workers • Annual number of newborns per 1000 persons • persons/km^2 • Percentage of buildings built more than 30 years ago among all buildings as of 2020 • Park and green area per person: Green area compared to the population in a dong (m^2/person)	• Statistics Korea Housing Census (2020) • Jeonju Data Portal Statistics Report (2020)
Individual parcel variables	**Building characteristics** • Built year • Parcel area • Building structure • Type of parcel • Road access conditions	• Reinforced concrete, block, wooden, steel frame, and others • Regular or irregular • Accessible to roads or not	• Seumteo Building Register (2020) • National Spatial Data Infrastructure Portal Open Market Cadastral Map (2020)
	Neighborhood environment characteristics (access) • Access to amenities • Presence of vacant properties • Presence of unwanted facilities	• Number of elementary schools, hypermarkets, hospitals, and welfare facilities located within the radius of 500 m per parcel (bus stop: 100 m) • Number of vacant properties within the radius of 500 m per parcel • Number of waste-treatment plants and wastewater discharge facilities within the radius of 500 m per parcel	• Jeonju Data Portal Statistics Report (2019)

Table 4. Characteristics of analyzed parcels.

Characteristics		Analyzed Parcels in Total		Vacant Properties		Parcels Except for Vacant Properties	
		Number	Percentage (%)	Number	Percentage (%)	Number	Percentage (%)
Building structure	Reinforced concrete	6786	21.3%	20	1.5%	6766	22.2%
	Block	19,120	60.1%	766	56.8%	18,354	60.3%
	Wooden	4916	15.5%	529	39.2%	4387	14.4%
	Steel frame	955	3.0%	16	1.2%	939	3.1%
	Others	33	0.1%	17	1.3%	16	0.1%
Type of parcel of land	Regular	7860	24.7%	475	35.2%	7385	24.2%
	Irregular	23,950	75.3%	873	64.8%	23,077	75.8%
Access	Not possible	8097	25.5%	720	53.4%	7377	24.2%
	Possible	23,713	74.5%	628	46.6%	23,085	75.8%
Construction year		37.0		52.8		36.3	
Parcel area (m^2)		288.4		447.8		281.4	

Basic statistics on variables also revealed that the percentage of aged buildings which were more than 30 years old among detached houses in Jeonju was 55.8%, and the average number of years since construction was $\geq$53—approximately 16.5 years older than non-vacant properties. As new buildings less than five years old accounted for only 1.2% of all buildings, a high percentage of old buildings is expected to worsen neighborhood environments and further exacerbate vacancies.

2.4. Methods

A multilevel logistic regression model calculates the probability that something will take place by considering other scales, and a logistic model is established based on exponential functions. The multilevel model is a statistical model that can be used to analyze data in hierarchical structures, where more than two units are included. This is also known as a hierarchical model. As vacant properties are locally affected by the characteristics assigned to the neighborhood level in addition to the individual building level, each effect may not be properly analyzed if the analysis is conducted at only one scale [20]. Because the multilayered model includes units at various scales when examining the effects of Dongs and the characteristics of buildings, as in this study, the model's advantage is that it can calculate the variance of the dependent variable by level and identify the effect at each level. Because the dependent variable—whether an individual parcel is vacant (vacant property = 1; non-vacant property = 0)—is a discrete variable, a multilevel logistic regression model with a logit link function, and not a linear multilayer model, was used. Because we analyzed environmental factors contributing to vacant properties by considering two spatial layers of individual parcels and neighborhood environments, a two-level multilayer model was established and used for analysis.

The logistic regression model equation is as follows.

$$P(Y_l = 1) = \frac{\exp(B_0 + B_1 \times X_l)}{1 + \exp(B_0 + B_1 \times X_l)} + e_l \tag{1}$$

$P(Y_l = 1)$ is the conditional probability that something will occur (vacant properties in this study). B_0 is the y-intercept; X_l is the predictor variable; B_1 is the coefficient for X_l, and e_l is the residual.

Because it is difficult to intuitively interpret the logistic regression model, transforming the s-shape curve in the exponential function into a linear line with the logit model as an alternative may make interpretation easier. The logit model equation is as follows:

$$Logit\ (odds) = B_0 + B_1 \times X_l \tag{2}$$

$Logit(odds)$ equals $Logit\left(\frac{P(Y_i=1)}{1-P(Y_i=1)}\right)$ and is interpreted as a logit scale for conditional probability. For instance, assuming X_i as the construction year and B_1 as 1.5, the odds ratio (OR) is $\exp(B_1) = \exp(1.5) \approx 4.5$. It is interpreted that an increase of one construction year increases the probability of vacancy by 4.5 times. In contrast, when B_1 is -1.5, it is $\exp(B_1) = \exp(-1.5) \approx 0.22$. It is $1/0.22 \approx 4.5$. It is therefore interpreted that a decrease of one construction year decreases the probability of vacancy by 4.5 times.

Before conducting a multilayer analysis, it must be validated by establishing an unrestricted model. An unrestricted model is a model in which only the dependent variable is input without any explanatory variable, and it consists only of the mean of the sample and the error term for each level. The sum of the error terms represents the total variance of the dependent variable, and the intra-class correlation can be obtained by calculating the percentage of the error term for each level. A high value means that the explanatory power is high at a certain level, and it is reasonable to analyze levels separately. In general, it is considered meaningful to use a multilayered model when it exceeds 5%. From the above process, we predicted the probability of vacancy for each detached house parcel, more objectively identified the spatial distribution of vacant properties using hotspot analysis, and confirmed the spatial distribution of vacant properties at risk of vacancy. The following equation was used:

$$ICC = \frac{var\left(u_{0j}\right)}{var\left(u_{0j}\right) + \left(\frac{\pi^2}{3}\right)} \tag{3}$$

where $var\left(u_{0j}\right)$ is the random intercept variance between Dongs.

The Stata/MP16 program was used for MLRM analysis. Then, hotspot analysis was conducted to determine where the vacant lands were statistically clustering by calculating the Getis-Ord Gi * of each parcel.

3. Results

3.1. Proportion of Variability in the Change of Vacancy

To determine whether a multilevel model is suitable for conducting the multilevel logistic regression model, we identified whether local characteristics contributed to an increase or decrease in vacancies between Dongs. Subsequently, a logistic regression analysis was conducted with 17 of the 18 variables extracted by considering previous studies, local characteristics, and data availability—except for the percentage of new buildings. New buildings built within the last five years were excluded from the analysis because they accounted for only 1.02% of the detached house parcels in Jeonju, which was considered insufficient for inclusion in the analysis.

Before analysis, we checked whether the variance in vacancy rate across Dongs was significant. As shown in Table 5, AIC and BIC were lower in the mixed effect model with a random intercept among Dongs than the standard logistic regression without a random intercept among Dongs. The difference was 755.06 with statistical significance at $p < 0.05$ in Chi-squared distributions. The results confirmed a local difference between Dongs in the effect on a vacancy increase or decrease, thus demonstrating the validity of the multilevel model.

Table 5. Comparison between models without a random intercept and those with a random intercept among Dongs.

	AIC	BIC	*p*-Value
Standard logistic regression without variables	9222.1	9281.7	0.000
Mixed effect model without variables	9096.6	9099.2	

We also examined variability in the probability of vacancy between Dongs with the ICC coefficient by considering only the random effect between Dongs without vacancy-related variables.

The ICC value indicates how much Dong scales compared to building scale account for the variations of vacancy. An analysis of local variance for the distribution of vacant properties, excluding the explanatory variable, with the above Equation (3) showed an intra-class correlation value in Jeonju of 0.206, which suggested that the variance in vacant properties in Jeonju is explained by 20.6% between Dongs (see Table 6). Since the ICC had high enough statistical reliability (over 5%), the results of volatility in the probability confirmed that the local characteristics of Jeonju greatly contributed to vacancy, and at the same time, proved that it is suitable to use the multilayered model. Therefore, we conducted a multilevel logistic regression model analysis to predict vacancies for each detached house parcel with the variables contributing to vacancy.

Table 6. Intraclass correlation coefficient value.

Level 1	Intra-Class Correlation	Std. Err.	[95% Conf. Interval]	
Dong Code	0.206	0.048	0.127	0.314

3.2. Multilevel Logistic Regression Model Analysis

Table 7 indicates the results of the multilevel logistic regression model analysis of variables that contribute to housing vacancies. The results of the analysis showed that 13 out of 20 variables had a statistically significant effect (95% significance) on vacancy. An analysis of the effect of each variable on vacant properties based on the estimated coefficient value showed that the probability of vacancy increased when the building structure was not concrete among other parcel and building characteristics. In particular, the variable that had the highest effect on vacant properties was the building structure and increasing the proportion of use of wooden structures by 1 (100%) increased the probability of the property being left vacant by 4.4 times compared to reinforced concrete structures. When detached houses and facilities in the neighborhood are old and poor and access to amenities decreases, preferences for the neighborhood decrease, which increases the probability of vacancy. Hence, it is necessary to continuously manage aged buildings with poor road access conditions.

Meanwhile, in terms of neighborhood environment characteristics, the presence of vacant properties or unwanted facilities around detached house parcels increased the probability of vacancy by about 1.17 times, while higher access to elementary schools and hypermarkets among other amenities led to a lower probability of vacancy. As such, vacant properties and unwanted facilities may harm the neighborhood, have a substantially negative effect on housing properties nearby, and eventually result in more vacant properties. High access to frequently used amenities, such as elementary schools and hypermarkets, also shows a negative relationship with vacant properties in our model, implying that the neighborhoods show less vacancies when anchor developments such as these are in place. The result indicates that such amenities are a positive attribute and result in decreased amounts of vacancy.

Table 7. The result of multilevel logistic regression analysis.

	Variables	Coefficient	SE	95% Coefficient Interval		Exp(Coefficient)	95% Coefficient Interval		Sig.
				Lower	Upper		Lower	Upper (%)	
	Intercept	−6.548	0.329	−7.193	−5.904	0.001	0.001	0.003	0.000
	Secondary industry	−0.862	0.603	−2.045	0.321	0.422	0.129	1.378	0.153
Dong Level	Fertility rate	0.019	0.007	0.005	0.034	1.019	1.005	1.034	0.009
	Population density	−0.000	0.000	−0.000	0.000	1.000	1.000	1.000	0.053
	Old building rate	0.702	0.287	0.140	1.264	2.018	1.151	3.539	0.014
	Park and green area	0.001	0.001	0.000	0.002	1.001	1.000	1.002	0.016
	Vacancy (500 m)	0.008	0.001	0.006	0.009	1.008	1.006	1.009	0.000
	School (500 m)	−0.192	0.066	−0.321	−0.064	0.825	0.726	0.938	0.003
	Hospital (500 m)	−0.087	0.079	−0.241	0.067	0.917	0.786	1.069	0.268
	Hypermarket (500 m)	−0.246	0.076	−0.395	−0.098	0.782	0.673	0.907	0.001
	Welfare facility (500 m)	−0.035	0.070	−0.174	0.103	0.965	0.841	1.108	0.615
Building Characteristics	Bus stop (100 m)	−0.102	0.069	−0.237	0.033	0.903	0.789	1.033	0.138
	Unwanted facility (500 m)	0.153	0.078	0.001	0.305	1.166	1.001	1.357	0.048
	Built year	0.025	0.002	0.021	0.028	1.025	1.021	1.029	0.000
	Parcel area	0.000	0.000	0.000	0.000	1.000	1.000	1.000	0.105
	Structure Block	1.353	0.240	0.884	1.823	3.871	2.421	6.190	0.000
	Structure Wood	1.477	0.256	0.975	1.980	4.382	2.652	7.239	0.000
	Structure Steel	1.324	0.348	0.642	2.005	3.758	1.901	7.429	0.000
	Structure Others	4.753	0.455	3.861	5.644	115.874	47.502	282.656	0.000
	Parcel type	0.075	0.067	−0.057	0.207	1.078	0.945	1.230	0.263
	Road access	0.649	0.063	0.526	0.773	1.914	1.692	2.166	0.000

Four out of the five Dong-level variables, except for the percentage of secondary industry workers, were found to have an effect on vacant properties. In particular, buildings over 30 years old had the highest effect, and population density and fertility rate had little effect. Unlike the common expectation that a higher percentage of secondary industry workers (mining, manufacturing and construction industry) will increase vacant properties, Jeonju showed opposite results. It seems that secondary industries, including manufacturing, still play a positive role in the local economy, as Jeonju is not yet a shrinking city experiencing a rapid population and economic decline. This phenomenon was confirmed to have a similar pattern in other studies on vacant properties in some growing cities [11,21,40]. Secondary industries take the raw materials produced by the primary sector and process them into manufactured goods and products (e.g., heavy and light manufacturing, food processing, oil refineries, energy production, etc.) and Jeonju's secondary industries still seem to be productive contributors to the local economy. It is important in the long term, however, that the city develops a method to phase out and reuse such industries as they become outdated or technological shifts require new facilities for existing services as to not increase their vacant land amount due to industrial life span dynamics.

3.3. Mean Multilevel Logistic Regression Model-Based Prediction of the Probability of Vacancy

Through The MLRM analysis with the above 20 variables, a prediction of the probability of vacancy for each detached house parcel in Jeonju predicted the probability of vacancy for total detached houses in Jeonju as 7.6%. Considering the current vacancy rate of 4.2% in Jeonju, we set detached house parcels with $\geq$5% in the probability of vacancy as areas at risk, those with $\geq$5 to <10% as areas at low risk, those with $\geq$10 to <15% as areas at moderate risk, and those with $\geq$15% as areas at high risk, and then confirmed their distribution. The results showed that 8723 parcels (27.4%) of the total parcels were currently vacant or $\geq$5% in the probability of vacancy, and 2537 parcels were currently used, but their probability of vacancy was $\geq$10% (Figure 4). Vacant properties are more likely to be distributed densely in a certain area than randomly spread across a wide area. In particular, the spatial density and spread of vacant properties accelerate a slum's growth into adjacent areas and facilities, leading to negative external effects such as deteriorating residential environments, risk of collapse, poor esthetics, increasing crime and disasters, falling housing prices, and increasing administrative costs.

To empirically analyze the above findings, we used a hotspot analysis to confirm the spatial distribution of parcels with a vacancy probability of $\geq$5% and identified the probability of vacancy by Dong (Figure 5). The results showed that the vacancy rate is expected to increase in all 35 Dongs. Furthermore, it is expected that areas with high vacancy rates will face a vacancy threat in the future. More specifically, the probability of vacancy in Area (1), an old city center, is expected to be the highest (24.7%). Areas (2) and (3), adjacent to the old city center, where construction activities are limited to prevent loss or damage to cultural heritage and protect the landscape in the area designated as a cultural heritage preservation site, are expected to have a higher probability of vacancy (14.5%). In contrast, Area (4), where real estate development is the most active and new large-scale commercial zones are being formed, had better residential environments than other areas, and its probability of vacancy is less than 1% (Table 8).

Table 8. Current and predicted vacancy probability by Dong.

	(1)	(2)	(3)	(4)
Current Vacancy (%)	0.14	0.06	0.06	0.00
Predicted Vacancy (%)	0.25	0.12	0.12	0.00

The result indicated that the spatial clustering of high-risk of vacancy has similar characteristics, such as structure type and year of construction. Thus, these physical conditions might have a stronger influence on the spatial cluster of vacancy than socioeconomic status.

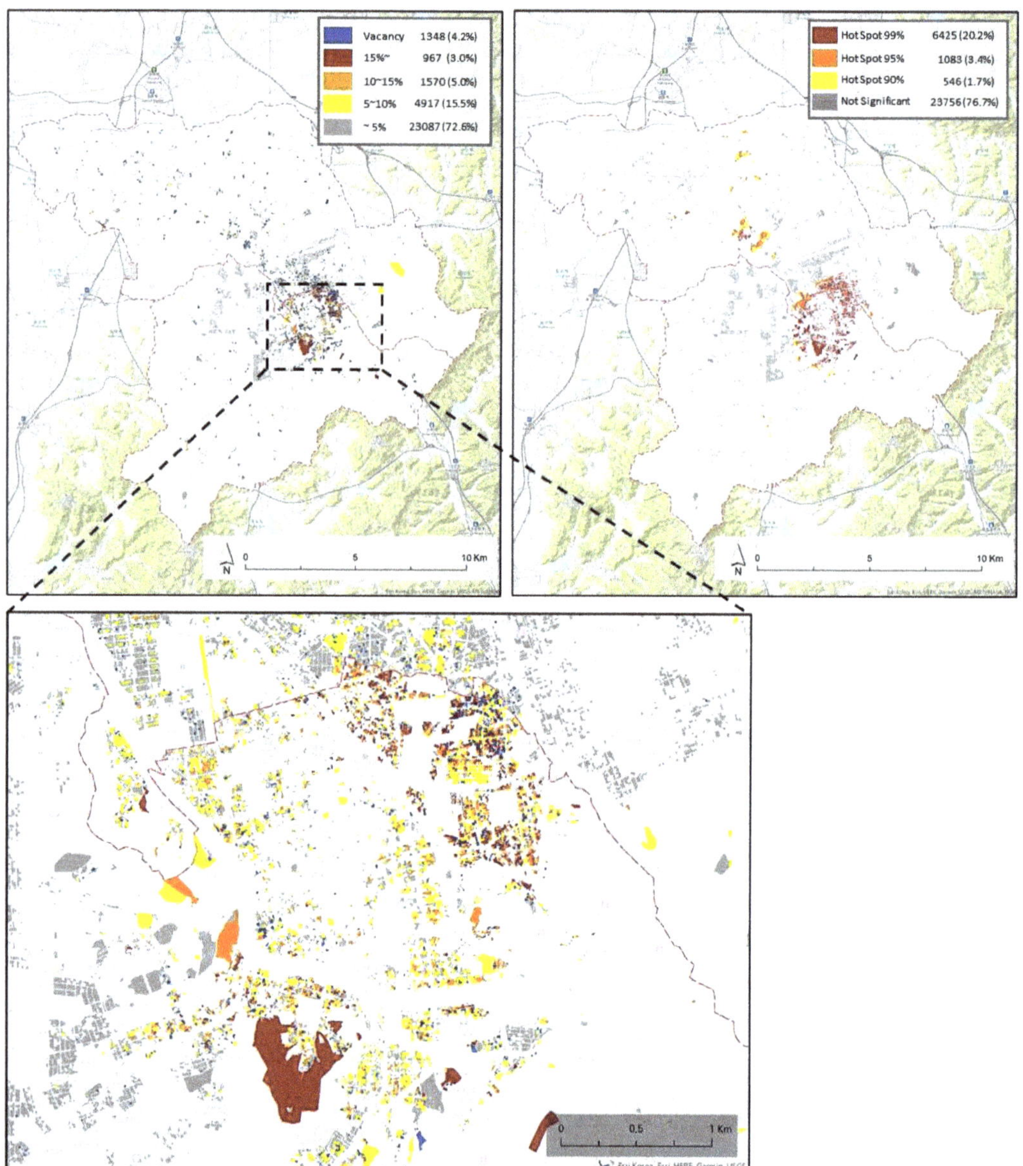

Figure 4. MLRM-based vacancy probability prediction (left) and hotspot analysis (right).

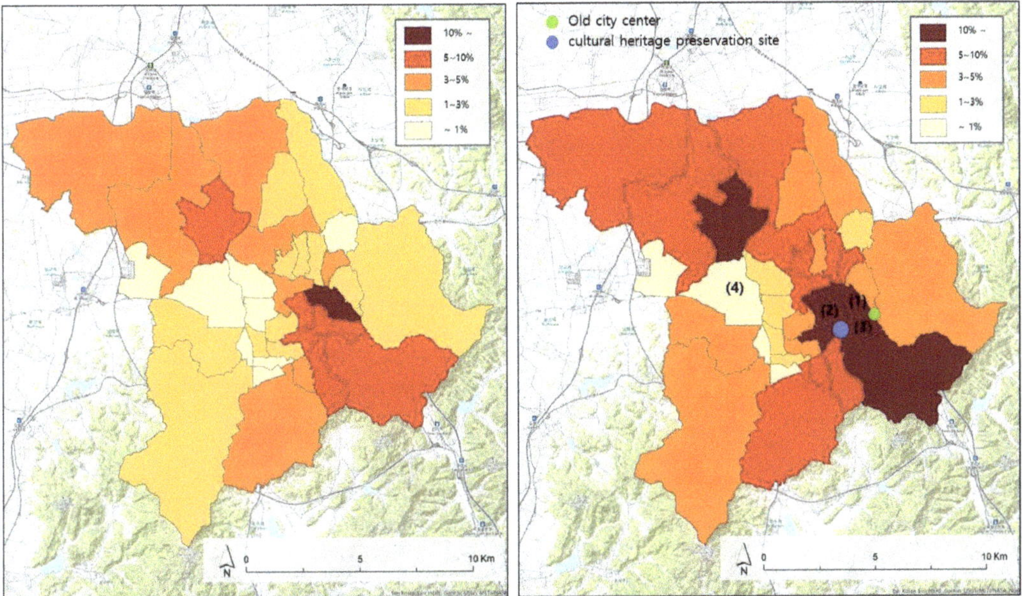

Figure 5. Dong (Census district) level current vacancy status (left) and vacancy prediction (right).

4. Conclusions

As vacant housing properties have recently emerged as a social issue, the central and local governments are making efforts to address this issue. To properly address the vacancy problem, it is more important than ever to establish a response strategy by first accurately identifying the contributors and characteristics of vacant properties, and then predicting areas at risk of vacancy in the future. We thus analyzed the causes and characteristics of vacant properties on a more local scale based on information on vacant properties in individual parcels in Jeonju. In addition, based on data for individual buildings, along with the local area's socioeconomic characteristics examined in previous studies, the analysis of contributors to vacant properties in this study included not only the unique physical characteristics of individual parcels but also the neighborhood environments as key explanatory variables.

Then, we predicted areas at risk of vacancy in the future and analyzed the spatial distribution of vacant properties. More importantly, to empirically examine the effect at different spatial levels, we established a multilayer model, conducted analysis, and overcame the limitations of previous studies that analyzed hierarchical data at one level. By predicting areas at risk of vacancy based on the analyzed results, we also provided a reference for using vacant properties tailored to the region and establishing urban regeneration strategies.

The following summary describes the results of this study. First, an analysis of factors contributing to vacant properties in Jeonju found that structure type and access to public facilities parks, schools, and hypermarkets, was related to vacant possibilities. This is consistent with the argument of Williams et al. (2019) [23], who showed the importance of schools and malls to thriving cities. It suggests that local governments need to improve access to public amenities.at Level 1 (individual parcels). In particular, wooden structures exhibited a high probability of vacancy. At Level 2 (Dong level), the fertility rate and old building rate were significantly associated with the vacant possibilities. This suggests that locally poor physical conditions and low rate of young people are related to the local economic recession to increase vacant possibilities at the local scale [41,42]. The results show that the local government should establish urban regeneration plans in Dongs with high fertility and old building rates. Second, a prediction of the probability of vacancy in Jeonju and a hotspot analysis showed that the old city center and areas with strict regulations on land use are expected to have a higher vacancy rate. Hence, it is necessary to prepare measures to repurpose vacant properties. Areas where vacant properties are concentrated and form a cluster require a careful analysis of urban spatial characteristics, including streets, blocks, and unwanted facilities, as well as the architectural characteristics of the properties.

5. Discussion

This study is significant in that it identified the distribution patterns of vacant properties, found various environmental factors contributing to vacant properties, and analyzed areas at high risk of vacancy in the future. However, as there are limitations in generalizing this study's results, an in-depth study would need to be conducted on the relationship between urban decline and vacant properties by region as well as the causes and characteristics of vacant properties. More notably, if time-series data were obtained with continuous follow-up of data on vacant properties, it would be possible to establish a more systematic management system for vacant housing properties.

Further, we conducted this study only on parcels, except for illegal buildings that did not have data on building characteristics. Nevertheless, as illegal buildings are actually old or highly likely to be vacant in the future, including them in the analysis would more accurately identify patterns in vacant properties and their changing effects on the properties. Finally, as this study's analysis was conducted only for vacant detached house properties, factors contributing to different types of vacant housing properties were not examined.

According to the results of this study, various effects from individual parcels, neighborhoods, and spatial ranges at the local level should be considered to solve issues related to excessive vacant properties. If neighborhood or local environments are in high-quality condition, it may be possible to solve such problems by upgrading individual vacant properties. If vacant properties are caused by complex effects at different spatial levels, it would be desirable to implement multidimensional measures such as improving the conditions of individual houses, upgrading neighborhood environments, and mitigating socioeconomic vulnerabilities at the local level. Further, it is also necessary to suggest different revitalization strategies based specifically on the regional characteristics of the area. In the central business districts, new anchor developments may spur regeneration in the highly vacant properties. However, it might be more helpful to promote place attachment and social relationships to increase the quality of life in residential areas by providing qualified community service facilities.

Despite the growing number of vacant properties in local small- and medium-sized cities owing to continuously outflowing and aging populations, most local governments continue to adhere to growth-oriented urban regeneration strategies that are unsupported by accurate empirical studies. Plans and projects to upgrade vacant properties should prioritize the identification of causes and patterns contributing to vacant properties based on an accurate survey of the actual status. Therefore, it is necessary to accurately recognize the causes of vacant properties, as identified in this study, and establish appropriate measures to reduce the number of vacant properties with policies. Accordingly, more efficient urban regeneration strategies can be established. This process must precede the identification of optimal alternatives and set project priorities when implementing urban regeneration projects.

Author Contributions: Conceptualization, J.L.; methodology, C.L.; writing—original draft preparation, J.L.; writing—review and editing, supervision, G.N.; funding acquisition, J.L. All authors have read and agreed to the published version of the manuscript.

Funding: This research was supported by the Ministry of Education of the Republic of Korea and the National Research Foundation of Korea (NRF-2019S1A5A8032562).

Data Availability Statement: Not applicable.

Acknowledgments: Not applicable.

Conflicts of Interest: The authors declare no conflict of interest.

References

1. Park, Y.; Newman, G.D.; Lee, J.-E.; Lee, S. Identifying and comparing vacant housing determinants across South Korean cities. *Appl. Geogr.* **2021**, *136*, 102566. [CrossRef]
2. Tarsitano, E.; Rosa, A.G.; Posca, C.; Petruzzi, G.; Mundo, M.; Colao, M. A sustainable urban regeneration project to protect biodiversity. *Urban Ecosyst.* **2021**, *24*, 827–844. [CrossRef]
3. Yoo, H.; Kwon, Y. Different Factors Affecting Vacant Housing According to Regional Characteristics in South Korea. *Sustainability* **2019**, *11*, 6913. [CrossRef]
4. Kato, H. Total Fertility Rate, Economic–Social Conditions, and Public Policies in OECD Countries. In *Macro-Econometric Analysis on Determinants of Fertility Behavior*; Springer: Singapore, 2021; pp. 51–76.
5. Kim, G.; Newman, G.; Jiang, B. Urban regeneration: Community engagement process for vacant land in declining cities. *Cities* **2020**, *102*, 102730. [CrossRef] [PubMed]
6. Pagano, M.A.; Bowman, A.O.M. *Vacant Land in Cities: An Urban Resource*; Brookings Institution, Center on Urban and Metropolitan Policy: Washington, DC, USA, 2000.
7. Olsen, A.K. Shrinking cities: Fuzzy concept or useful framework? *Berkeley Plan. J.* **2013**, *26*, 107–132. [CrossRef]
8. Korea, S. 2017 Population and housing census. In *Housing Census*; Statistics Korea: Daejeon, Korea, 2018.
9. Pagano, M.; Bowman, A. Vacant land as opportunity and challenge. In *Recycling the City: The Use and Reuse of Urban Land*; Lincoln Institute: Cambridge, MA, USA, 2004; pp. 15–32.
10. Lee, J.; Newman, G. Forecasting Urban Vacancy Dynamics in a Shrinking City: A Land Transformation Model. *ISPRS Int. J. Geo-Inf.* **2017**, *6*, 124. [CrossRef]
11. Lee, J.; Park, Y.; Kim, H.W. Evaluation of Local Comprehensive Plans to Vacancy Issue in a Growing and Shrinking City. *Sustainability* **2019**, *11*, 4966. [CrossRef]
12. Scafidi, B.P.; Schill, M.H.; Wachter, S.M.; Culhane, D.P. An economic analysis of housing abandonment. *J. Hous. Econ.* **1998**, *7*, 287–303. [CrossRef]
13. Hillier, A.E.; Culhane, D.P.; Smith, T.E.; Tomlin, C.D. Predicting housing abandonment with the Philadelphia neighborhood information system. *J. Urban Aff.* **2003**, *25*, 91–106. [CrossRef]
14. Bassett, E.M.; Schweitzer, J.; Panken, S. *Understanding Housing Abandonment and Owner Decision-Making in Flint, Michigan: An Exploratory Analysis*; Lincoln Institute of Land Policy Working Paper; Lincoln Institute of Land Policy Information Services: Cambridge, MA, USA, 2006; pp. 1–62.
15. Silverman, R.M.; Yin, L.; Patterson, K.L. Dawn of the dead city: An exploratory analysis of vacant addresses in Buffalo, NY 2008–2010. *J. Urban Aff.* **2013**, *35*, 131–152. [CrossRef]
16. Morckel, V.C. Empty neighborhoods: Using constructs to predict the probability of housing abandonment. *Hous. Policy Debate* **2013**, *23*, 469–496. [CrossRef]
17. Keenan, P.; Lowe, S.; Spencer, S. Housing abandonment in inner cities-the politics of low demand for housing. *Hous. Stud.* **1999**, *14*, 703–716. [CrossRef]
18. Mallach, A. *Bringing Buildings Back: From Abandoned Properties to Community Assets: A Guidebook for Policymakers and Practitioners*; Rutgers University Press: New Brunswick, NJ, USA, 2006.
19. Sternlieb, G.; Burchell, R.W.; Hughes, J.W.; James, F.J. Housing abandonment in the urban core. *J. Am. Inst. Plan.* **1974**, *40*, 321–332. [CrossRef]
20. Accordino, J.; Johnson, G.T. Addressing the vacant and abandoned property problem. *J. Urban Aff.* **2000**, *22*, 301–315. [CrossRef]
21. Newman, G.; Lee, J.; Berke, P. Using the land transformation model to forecast vacant land. *J. Land Use Sci.* **2016**, *11*, 450–475. [CrossRef]
22. Zou, S.; Wang, L. Individual vacant house detection in very-high-resolution remote sensing images. *Ann. Am. Assoc. Geogr.* **2019**, *110*, 449–461.
23. Williams, S.; Xu, W.; Bin Tan, S.; Foster, M.J.; Chen, C. Ghost cities of China: Identifying urban vacancy through social media data. *Cities* **2019**, *94*, 275–285. [CrossRef]
24. Efthymiou, D.; Antoniou, C. How do transport infrastructure and policies affect house prices and rents? *Evidence from Athens, Greece. Transp. Res. Part A Policy Pr.* **2013**, *52*, 1–22. [CrossRef]
25. Palm, M.; Gregor, B.; Wang, H.; McMullen, B.S. The trade-offs between population density and households' transportation-housing costs. *Transp. Policy* **2014**, *36*, 160–172. [CrossRef]
26. Song, Z.; Cao, M.; Han, T.; Hickman, R. Public transport accessibility and housing value uplift: Evidence from the Docklands light railway in London. *Case Stud. Transp. Policy* **2019**, *7*, 607–616. [CrossRef]
27. Jeon, Y.; Kim, S. The causes and characteristics of housing abandonment in an inner-city neighborhood—Focused on the Sungui-dong area, Nam-gu, Incheon. *J. Urban Des. Inst. Korea Urban Des.* **2016**, *17*, 83–100.
28. Park, J.-I. A Multilevel Model Approach for Assessing the Effects of House and Neighborhood Characteristics on Housing Vacancy: A Case of Daegu, South Korea. *Sustainability* **2019**, *11*, 2515. [CrossRef]

29. Masuda, H. *Chihō Shōmetsu [The Extinction of Rural Areas]*; Chūkō Shinsho: Tokyo, Japan, 2014.
30. Lee, S.-H. Local Regions' Extinction in South Korea 2018: Focusing on the Trend of 2013~2018 and Population Movement in Non-Capital Region Areas. *Employ. Trend Brief* **2018**, *2018*, 2–21.
31. Lee, S. *Seven Analysis on the Local Extinction in Korea, Regional Employment Trend Brief*; Korea Employment Information Service: Seoul, Korea, 2016.
32. Shah, I.H.; Dong, L.; Park, H.-S. Tracking urban sustainability transition: An eco-efficiency analysis on eco-industrial development in Ulsan, Korea. *J. Clean. Prod.* **2020**, *262*, 121286. [CrossRef]
33. Jo, Y.-W.; Choi, Y.-B.; Park, C. Exploring Spatial Distribution of Empty Houses and Vacant Land Due to Population Decrease in Mokpo. *J. Korean Soc. Environ. Restor. Technol.* **2020**, *23*, 33–47.
34. Osland, L.; Östh, J.; Nordvik, V. House price valuation of environmental amenities: An application of GIS-derived data. *Reg. Sci. Policy Pr.* **2020**, 1–21. [CrossRef]
35. Kaido, K. Urban densities, quality of life and local facility accessibility in principal Japanese cities. In *Future forms and Design for Sustainable Cities*; Routledge: London, UK, 2006; pp. 320–347.
36. Niu, S.; Hu, A.; Shen, Z.; Lau, S.S.Y.; Gan, X. Study on land use characteristics of rail transit TOD sites in new towns—taking Singapore as an example. *J. Asian Arch. Build. Eng.* **2019**, *18*, 16–27. [CrossRef]
37. Cozens, P.; Tarca, M. Exploring housing maintenance and vacancy in Western Australia: Perceptions of crime and crime prevention through environmental design (CPTED). *Prop. Manag.* **2016**, *34*, 199–220. [CrossRef]
38. Pallagst, K. From urban shrinkage to urban qualities? *J. Urban Des.* **2019**, *24*, 68–70. [CrossRef]
39. Schilling, J.; Logan, J. Greening the rust belt: A green infrastructure model for right sizing America's shrinking cities. *J. Am. Plan. Assoc.* **2008**, *74*, 451–466. [CrossRef]
40. Lee, J.; Newman, G.; Park, Y. A Comparison of Vacancy Dynamics between Growing and Shrinking Cities Using the Land Transformation Model. *Sustainability* **2018**, *10*, 1513. [CrossRef] [PubMed]
41. Szafrańska, E.; De Lille, L.C.; Kazimierczak, J. Urban shrinkage and housing in a post-socialist city: Relationship between the demographic evolution and housing development in Łódź, Poland. *Neth. J. Hous. Built Environ.* **2019**, *34*, 441–464. [CrossRef]
42. Esther Yakubu, I.; Egbelakin, T.; Dizhur, D.; Ingham, J.; Sungho Park, K.; Phipps, R. Why are older inner-city buildings vacant? *Implications for town centre regeneration. J. Urban Regen. Renew.* **2017**, *11*, 44–59.

sustainability

Article

Technology for Predicting Particulate Matter Emissions at Construction Sites in South Korea

Jihwan Yang [1], Sungho Tae [2,*] and Hyunsik Kim [3,*]

1 Department of Living and Built Environment Research, Seoul Institute of Technology, Mapo-gu, Seoul 03909, Korea; jhyang@sit.re.kr
2 School of Architecture & Architectural Engineering, Hanyang University, Sangnok-gu, Ansan 15588, Korea
3 Department of Architectural Engineering, Hanyang University, Sangnok-gu, Ansan 15588, Korea
* Correspondence: jnb55@hanyang.ac.kr (S.T.); visionysj@gmail.com (H.K.);
   Tel.: +82-31-400-5187 (S.T.); +82-31-436-8080 (H.K.)

Abstract: In recent years, particulate matter (PM) has emerged as a major social issue in various industries, particularly in East Asia. PM not only causes various environmental, social, and economic problems but also has a large impact on public health. Thus, there is an urgent requirement for reducing PM emissions. In South Korea, the PM generated at construction sites in urban areas directly or indirectly causes various environmental problems in surrounding areas. Construction sites are considered a major source of PM that must be managed at the national level. Therefore, this study aims to develop a technology for predicting PM emissions at construction sites. First, the major sources of PM at construction sites are determined. Then, PM emission factors are calculated for each source. Furthermore, an algorithm is developed for calculating PM emissions on the basis of an emission factor database, and a system is built for predicting PM emissions at construction sites. The reliability of the proposed technology is evaluated through a case study. The technology is expected to be used for predicting potential PM emissions at construction sites before the start of construction.

Keywords: construction site; particulate matter emissions; emission factor; prediction technology

Citation: Yang, J.; Tae, S.; Kim, H. Technology for Predicting Particulate Matter Emissions at Construction Sites in South Korea. *Sustainability* **2021**, *13*, 13792. https://doi.org/10.3390/su132413792

Academic Editor: Hyun-Woo Kim

Received: 25 October 2021
Accepted: 9 December 2021
Published: 14 December 2021

Publisher's Note: MDPI stays neutral with regard to jurisdictional claims in published maps and institutional affiliations.

1. Introduction

The World Health Organization has recently classified particulate matter (PM) as a class 1 carcinogen because research has shown that it may cause lung, cardiovascular, and respiratory diseases. Thus, considerable effort is being made to reduce PM emissions in various industries worldwide [1–9]. In East Asian countries, particularly South Korea and China, the emission of large amounts of PM due to radical industrialization is emerging as a major social problem [10–15]. According to the statistical data of the Ministry of Environment (ME) of South Korea, fugitive emissions account for approximately 50% of PM10 and smaller PM emissions in South Korea. Furthermore, the fugitive emissions caused by construction work are the largest at 33% [16,17].

PM is generated during construction mainly by the movement of construction equipment and construction activities. Accordingly, there is an urgent requirement for research on various PM emission sources associated with construction equipment [18]. Currently, there is no standard for regulating PM emissions at construction sites in South Korea and no method for calculating PM emissions to set such a standard [19]. The National Institute of Environmental Research (NIER) of South Korea uses the method provided by the US Environmental Protection Agency (EPA) to calculate the amount of PM10 in fugitive dust at construction sites using an equation [20]. However, the equation calculates PM10 using only the construction period and the size of a construction area, and it cannot reflect the various construction conditions that generate PM.

In South Korea, PM at construction sites is managed on the basis of the emission concentration through "PM Emergency Reduction Measures" [21]. However, this is a passive method that is applied after a high concentration of PM has already been generated. Therefore, a method to preventively manage PM emissions by controlling the quantity of emissions is needed for accurate PM management. As the South Korean government has recognized the need for quantitative management of dust emission, the "Business Site Total Air Pollution Management System" was implemented nationwide in South Korea in 2020. This is a preemptive system for managing air pollutants by presenting quantitative emission standards in advance. It has realized an active reduction in air pollution by setting the total amount of allowed emissions and then assigning an amount to each business site to maintain pollutants within the allotted range [22]. However, this system only regulates the business site emissions and does not include construction sites as management targets. The South Korean government continuously discusses the importance of managing construction site emissions; however, there is no clear standard to evaluate PM emissions generated from construction sites. Therefore, it is necessary to develop a systematic method for calculating PM emissions at construction sites quantitively by developing emission factors and calculating methods to expand the national air pollution management system to construction sites in the future [23].

Therefore, this study develops a technology to predict PM emissions at construction sites by calculating the emission factors for PM10 and PM2.5 for major PM emission sources. PM emissions are calculated on the basis of an emission factor database (DB). Furthermore, a system is developed for predicting PM emissions. The reliability of the prediction technology is examined through a case study.

2. Materials and Methods

The PM emission factors at construction sites were calculated and used to construct a DB. Then, this DB was utilized to develop a method for calculating PM emissions. Finally, an algorithm was developed to predict PM emissions at actual construction sites. The PM emission factors were calculated by identifying the major sources of PM emissions and construction activities. The emission factor DB consisted of direct and fugitive emission factors for each construction activity and emission source. In addition, emission scenarios and calculation methods were proposed for each emission source. Finally, a system for predicting PM emissions at construction sites was developed on the basis of the emission factor DB and calculation method. Figure 1 shows the research method of this study.

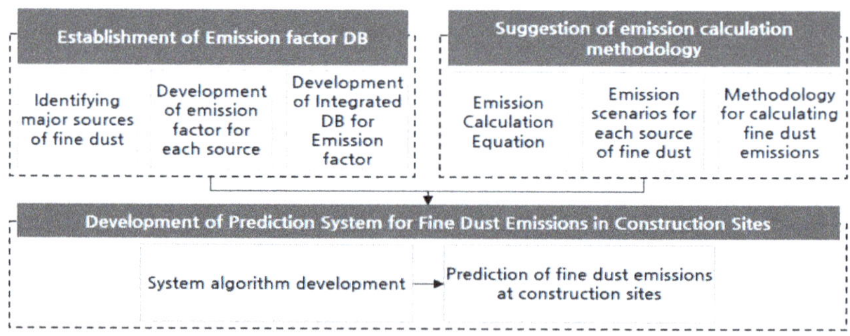

Figure 1. Research framework and methods.

2.1. Major PM Emission Sources at Construction Sites

Earthworks are considered the major source of PM emissions at construction sites [24]. In addition, according to US EPA AP-42 and the ME of South Korea, the movement of construction equipment used for civil works and construction activities is a major source of PM emissions at construction sites [20,25]. Figure 2 shows the major PM emission sources selected in this study. These sources were selected on the basis of the construction-related air pollutant emission sources defined by the US EPA and ME of South Korea [17,20,25–27].

Among the large category of air pollutant emission sources defined in South Korea, the categories of on-road and nonroad mobile pollution sources were selected. The major dust generating sources defined in AP-42 (Heavy Construction Operations) of US EPA match with the nine types of construction equipment and four types of material transport equipment in the small category of South Korean air pollutant emission sources. As a result, 13 major PM emission sources were selected in this study.

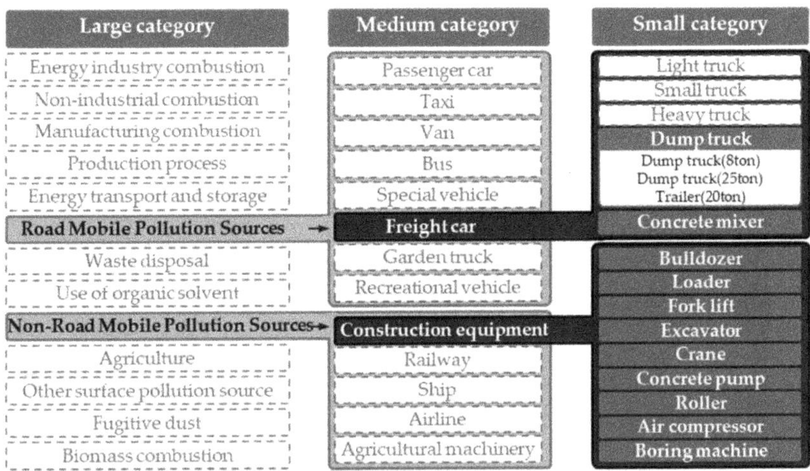

Figure 2. Major particulate matter (PM) generating sources in construction sites.

2.2. PM Emission Factor DB for Construction Sites

Two types of PM emissions due to construction equipment were considered. The first type was direct emission, in which PM is released into the air through fuel combustion in construction equipment. The second type was fugitive emission, in which PM is released by the construction activities of construction equipment. Different methods were used to calculate the emission factors for each type of emission.

For direct emission, the air pollutant emission factor data presented by the NIER of South Korea were applied mutatis mutandis [26]. The emission factors for dump trucks and trailers were directly specified in the freight car category of on-road mobile pollution sources. The emission factors for the nine types of construction equipment were defined according to the rated output of each equipment type [20]. This study directly calculated the emission factors using the average rated output data for construction equipment manufactured in South Korea in accordance with the National Air Pollutant Emission Calculation Method Manual of South Korea. Table 1 shows the direct emission factors for PM10 and PM2.5 for different types of construction equipment.

For fugitive emission, the emission factors were calculated according to US EPA AP-42 [27–29]. However, forklifts, concrete pumps, and air compressors were excluded because the definition of fugitive emission activities was unclear. Table 2 shows the equations for calculating the emission factors for the construction activities of construction equipment. The information required for calculating the emission factors, such as the silt content, moisture content, and mean wind speed, was obtained from South Korean literature (GRI, 2019) [25].

Table 1. Direct emission factors for construction equipment according to rated output.

Equipment	Rated Power (kW) [20]	Emission Factor (kg/kWh) [27]	
		PM10	**PM2.5**
Bulldozer	114	0.00022000	0.00020240
Loader	100	0.00022000	0.00020240
Forklift	56	0.00028000	0.00025760
Excavator	85	0.00019000	0.00017480
Crane	175	0.00012000	0.00011040
Roller	70	0.00034000	0.00031280
Air compressor	201	0.00010000	0.00009200
Concrete pump	199	0.00002000	0.00001840
Boring machine	106	0.00012000	0.00011040
Dump truck, trailer	Average vehicle speed V = 20 km/h	4.30×10^{-4} (kg/km)	3.96×10^{-4} (kg/km)

Table 2. Formulae for calculating fugitive emission factors for construction equipment.

Equipment	Construction Work [27]	Emission Factor Calculation [28,29]	
		PM10	**PM2.5**
Bulldozer	Bulldozing	$0.75[0.45(s)^{1.5}/(M)^{1.4}]$ (kg/h)	$0.105[2.6(s)^{1.2}/(M)^{1.3}]$ (kg/h)
Loader	Loading material	$0.35(0.0016) \times [(U/2.2)^{1.3}/(M/2)^{1.4}]$ (kg/Mg)	$0.053(0.0016) \times [(U/2.2)^{1.3}/(M/2)^{1.4}]$ (kg/Mg)
Forklift	Vehicular traffic	No fugitive emission factor	
Excavator	Power shovel	0.018(00005/0.001)	0.018(0.0001/0.001)
	Loading material	$0.35(0.0016) \times [(U/2.2)^{1.3}/(M/2)^{1.4}]$ (kg/Mg)	$0.053(0.0016) \times [(U/2.2)^{1.3}/(M/2)^{1.4}]$ (kg/Mg)
Crane	Vehicular traffic	$1.5(s/12)^{0.9}(W/3)^{0.45} \times 0.2819$ (kg/VKT)	$0.15(s/12)^{0.9}(W/3)^{0.45} \times 0.2819$ (kg/VKT)
Concrete pump	Pumping	No fugitive emission factor	
Roller	Compacting	$0.75[0.45(s)^{1.5}/(M)^{1.4}]$ (kg/h)	$0.105[2.6(s)^{1.2}/(M)^{1.3}]$ (kg/h)
Air compressor	Painting	No fugitive emission factor	
Boring machine	Drilling	$0.59 \times (0.16/1.3)$ (kg/hole)	$0.59 \times (0.16/1.3) \times 0.2$ (kg/hole)
Dump truck	Dumping	$0.001 \times (0.001/0.002)$ (kg/ton)	$0.001 \times (0.001/0.002) \times 0.2$ (kg/ton)
	Vehicular traffic (25 tons)	$1.5(s/12)^{0.9}(W/3)^{0.45} \times 0.2819$ (kg/VKT)	$0.15(s/12)^{0.9}(W/3)^{0.45} \times 0.2819$ (kg/VKT)
	Vehicular traffic (8 tons)	$1.5(s/12)^{0.9}(W/3)^{0.45} \times 0.2819$ (kg/VKT)	$0.15(s/12)^{0.9}(W/3)^{0.45} \times 0.2819$ (kg/VKT)
Trailer	Vehicular traffic (20 ton trailer)	$1.5(s/12)^{0.9}(W/3)^{0.45} \times 0.2819$ (kg/VKT)	$0.15(s/12)^{0.9}(W/3)^{0.45} \times 0.2819$ (kg/VKT)
Concrete mixer	Vehicular traffic (15 tons)	$1.5(s/12)^{0.9}(W/3)^{0.45} \times 0.2819$ (kg/VKT)	$0.15(s/12)^{0.9}(W/3)^{0.45} \times 0.2819$ (kg/VKT)

s: silt content (%); M: moisture content (%); U: mean wind speed (m/s); W: vehicle weight (ton). s = 9; M = 12; U = 3.65; W = various [25].

However, according to the Korean national air pollutant emission factors (NIER, 2015) which are used for calculating the annual emission in Korea, fugitive emissions are calculated using Equation (1) and the emission factors in Table 3 [26].

$$E = \sum A \times P \times EF \tag{1}$$

where E denotes the total fugitive emissions (kg/year), A denotes the annual construction area (m^2), P denotes the annual earthwork construction period (month/year), and EF denotes emission factor (kg/m^2/month).

Table 3. Emission factors (kg/m^2/month) for Equation (1) [26].

Construction Type		PM10	PM2.5
Residential	House	0.0072	0.00072
	Apartment	0.0247	0.00247
Nonresidential		0.0426	0.00426
Road construction		0.0941	0.00941

The emission factors were derived from a NIER study (NIER, 2008) to select the most appropriate emission factors for construction sites among the emission factors developed by the US EPA [30]. According to this research study, the fugitive emission factors for each construction equipment were calculated using AP-42 as in this study. However, the types of construction equipment considered in the research were limited compared to this study, which implies that the fugitive emission factor database of this study is more appropriate for calculating the fugitive emissions of construction sites compared to Equation (1). The comparison of construction equipment considered in both studies is shown in Table 4.

Table 4. Construction equipment considered for fugitive emission factor development.

Studies	Construction Equipment
NIER, 2008 [30]	Bulldozer, excavator, roller, dump truck
This study	Bulldozer, loader, excavator, crane, concrete pump, roller, compressor, boring machine, forklift, concrete mixer truck, dump truck, trailer

2.3. Calculation Method for PM Emissions at Construction Sites

The PM emissions at construction sites in South Korea were calculated using Equation (2) of US EPA AP-42. This equation requires the activity rate, emission factor, and overall emission reduction efficiency, which quantify the degree and scope of construction activities. These variables should be presented as applicable figures in South Korea. The activity rate applicable to South Korea was obtained by deriving the construction activity scenarios for each type of construction equipment using the Standard of Construction Estimate [31]. In particular, the activity rates for bulldozers and rollers were derived by selecting the most typical usage of each equipment type, and other scenarios were derived by following the use of each equipment type described in the standard. Table 5 shows the construction activity scenarios for the construction equipment. In addition, the emission factor applicable to South Korea was obtained using the emission factor DB described in Section 2.2.

$$E = A \times EF \times \left(\frac{1 - ER}{100} \right) \tag{2}$$

E, A, EF, and ER denote the emissions, activity rate, emission factor, and overall emission reduction efficiency, respectively. However, ER in this study was considered 0, as there is no standard for ER.

factors. The direct emissions of PM due to the unloading of dump trucks were considered to be negligible and excluded from this study.

3.2. System for Predicting PM Emissions at Construction Sites

Figure 3 shows the components and algorithm of the system for predicting PM emissions at construction sites. The total PM emissions at construction sites were the sum of the direct and fugitive emissions caused by construction equipment, which were calculated as described in Section 3.1. An Excel-based system was developed for predicting PM emissions at construction sites. Figure 4 shows the information input screen of the system, where a user enters a design statement and general information about a construction project, and a sheet for displaying the evaluation result. The information input sheet was divided into two parts for entering the general information and design statement. The general information part was configured to input general site information such as the name of the construction project, construction period, site area, gross floor area, underground depth, and ground height. The design statement part was configured to input the amount of earthwork and construction material, which determined the amount of activity of construction equipment, on the basis of design documents and detailed statements.

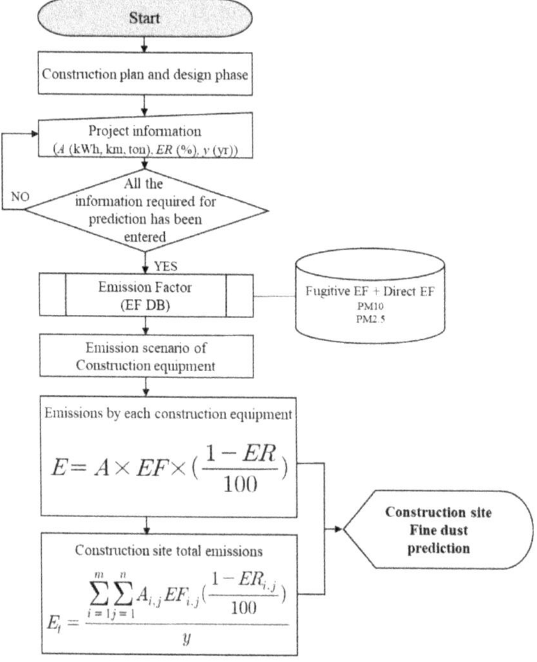

$$E = A \times EF \times \left(\frac{1 - ER}{100}\right)$$

$$E_t = \frac{\sum_{i=1}^{m}\sum_{j=1}^{n} A_{i,j} EF_{i,j}\left(\frac{1 - ER_{i,j}}{100}\right)}{y}$$

Figure 3. Algorithm of system for predicting PM emissions at construction sites (*EF*: emission factor).

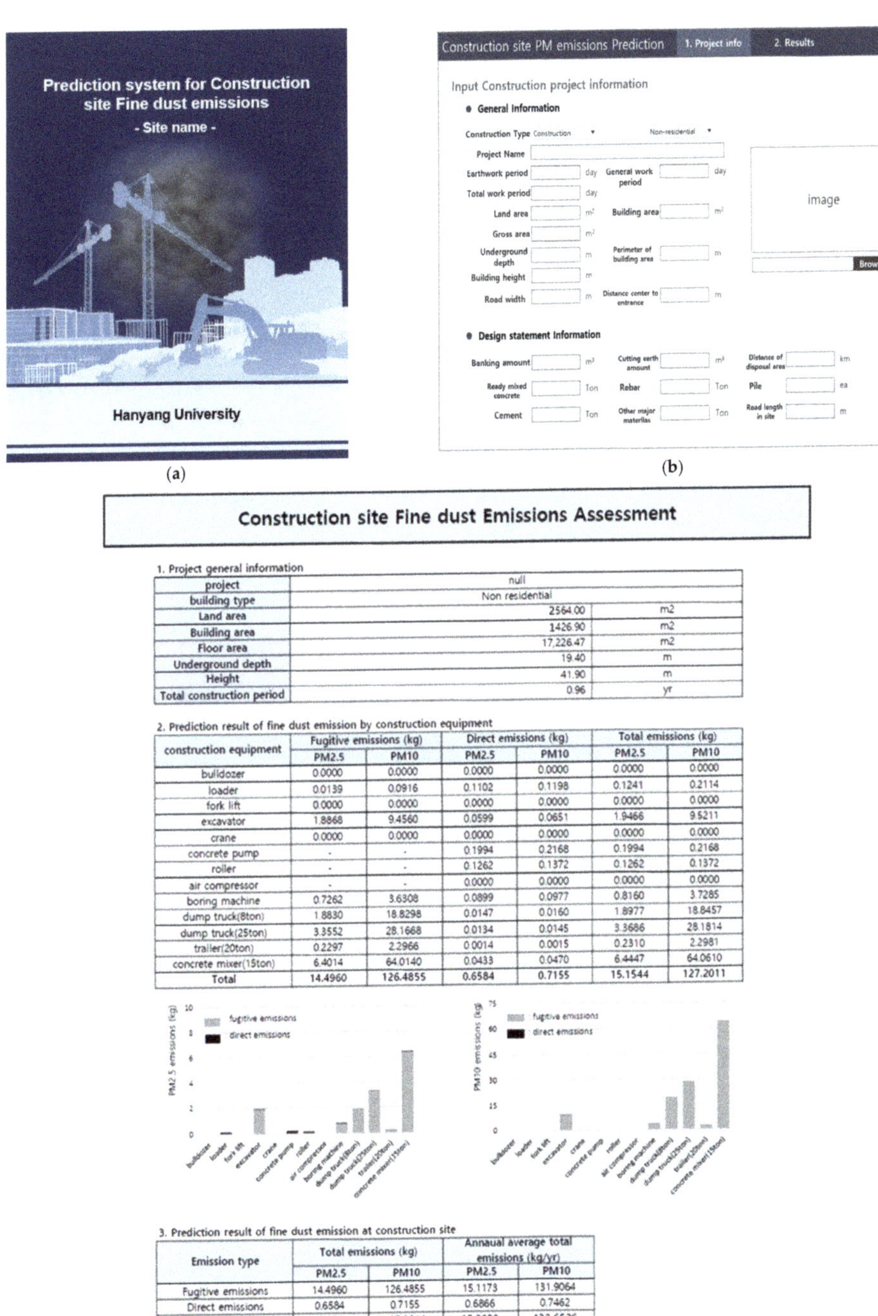

Figure 4. System for predicting PM emissions at construction sites. (**a**) Main screen, (**b**) Information input sheet, (**c**) Assessment result sheet.

4. Case Study

4.1. Overview of Case Study

The applicability of the developed system was examined by predicting PM emissions using actual construction site information, as shown in Table 7. This information was obtained from the design outline, elevation drawing, floor plan, and design details of the building. The duration of earthworks and general works and the number of concrete piles were not available; these were assumed based on the scale of construction. The amount of work was calculated according to the PM emission scenario at the construction site in the prediction system. Then, the final PM emissions were predicted using the calculated amount of work and established emission factor DB.

Table 7. Information of the construction site to be evaluated (PM: particulate matter).

	Project Name	**Case Study for PM Emission Prediction at Construction Site**		**Building Use**	**Nonresidential**
General information	Civil construction period	200 days	General construction period	150 days	Total construction period	350 days
	Land area	2564 m^2	Building area	1426.90 m^2	Gross floor area	17,226.47 m^2
	Underground depth	19.40 m	Above-ground height	41.90 m	Building area perimeter	186.70 m
	Road width on site	8.00 m	Average moving distance	39.26 m	Construction type	Building construction
Statement information	Fill amount	6845.00 m^3	Cut amount	1176.00 m^3	Distance to waste soil dumpsite	0.50 km
	Amount of ready-mixed concrete work	20,879.60 tons	Amount of rebar work	868.67 tons	Pile hole	50 holes
	Amount of cement work	1755.61 tons	Other major materials	22,024.70 tons	Road distance on site	2000 m

4.2. Results of Case Study

Table 8 shows the evaluation results of PM emissions by construction equipment. The PM10 emissions by the concrete mixer truck were the largest (50.36% of the total PM10 emissions), followed by the 25-ton dump truck, 8-ton dump truck, and excavator. The PM10 emissions from these types of equipment accounted for approximately 95% of the total emissions at the construction site. The PM2.5 emissions showed a similar trend, with a few differences. The concrete mixer truck generated the largest amount of PM2.5 emissions. Nevertheless, they accounted for only 42.53% of the total emissions. Furthermore, the PM2.5 emissions decreased in the order of the 25-ton dump truck, excavator, and 8-ton dump truck. Table 9 shows the predicted total emissions and Figure 5 shows the predicted PM2.5 and PM10 emissions for construction equipment. The concrete mixer truck and dump trucks (25 tons and 8 tons) emitted high amounts of PM. Thus, a plan for reducing PM emissions should focus on these types of equipment.

Table 8. PM emissions by construction equipment.

Construction Equipment	Fugitive Emissions (kg)		Direct Emissions (kg)		Total Emissions (kg)	
	PM2.5	PM10	PM2.5	PM10	PM2.5	PM10
Bulldozer	0.00×10^0	0.00×10^0	0.00×10^0	0.00×10^0	0.00×10^0	0.00×10^0
Loader	1.39×10^{-2}	9.16×10^{-2}	1.10×10^{-1}	1.20×10^{-1}	1.24×10^{-1}	2.11×10^{-1}
Forklift	0.00×10^0	0.00×10^0	0.00×10^0	0.00×10^0	0.00×10^0	0.00×10^0
Excavator	1.89×10^0	9.46×10^0	5.99×10^{-2}	6.51×10^{-2}	1.95×10^0	9.52×10^0
Crane	0.00×10^0	0.00×10^0	0.00×10^0	0.00×10^0	0.00×10^0	0.00×10^0
Concrete pump	-	-	1.99×10^{-1}	2.17×10^{-1}	1.99×10^{-1}	2.17×10^{-1}
Roller	-	-	1.26×10^{-1}	1.37×10^{-1}	1.26×10^{-1}	1.37×10^{-1}
Air compressor	-	-	0.00×10^0	0.00×10^0	0.00×10^0	0.00×10^0
Boring machine	7.26×10^{-1}	3.63×10^0	8.99×10^{-2}	9.77×10^{-2}	8.16×10^{-1}	3.73×10^0
Dump truck (8 ton)	1.88×10^0	1.88×10^1	1.47×10^{-2}	1.60×10^{-2}	1.90×10^0	1.88×10^1
Dump truck (25 tons)	3.36×10^0	2.82×10^1	1.34×10^{-2}	1.45×10^{-2}	3.37×10^0	2.82×10^1
Trailer (20 tons)	2.30×10^{-1}	2.30×10^0	1.35×10^{-3}	1.47×10^{-3}	2.31×10^{-1}	2.30×10^0
Concrete mixer (15 tons)	6.40×10^0	6.40×10^1	4.33×10^{-2}	4.70×10^{-2}	6.44×10^0	6.41×10^1
Total	1.45×10^1	1.26×10^2	6.58×10^{-1}	7.16×10^{-1}	1.52×10^1	1.27×10^2

Table 9. Results obtained using system for predicting PM emissions at construction sites.

Emission Type	Total Emissions (kg)		Annual Average Total Emissions (kg/year)	
	PM2.5	PM10	PM2.5	PM10
Fugitive emissions	1.45×10^1	1.26×10^2	1.51×10^1	1.32×10^2
Direct emissions	6.58×10^{-1}	7.16×10^{-1}	6.87×10^{-1}	7.46×10^{-1}
Total	1.52×10^1	1.27×10^2	1.58×10^1	1.33×10^2

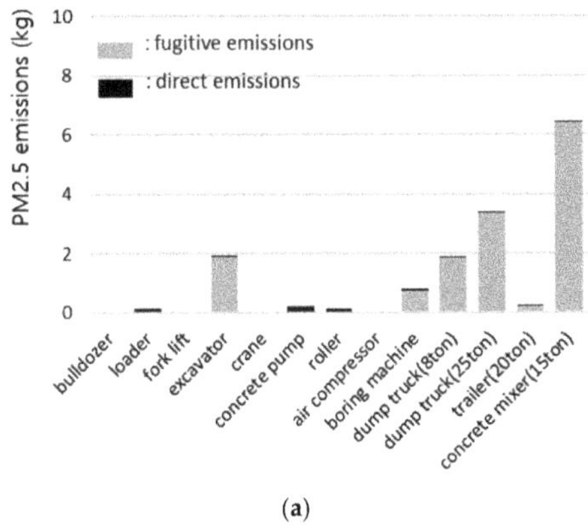

(**a**)

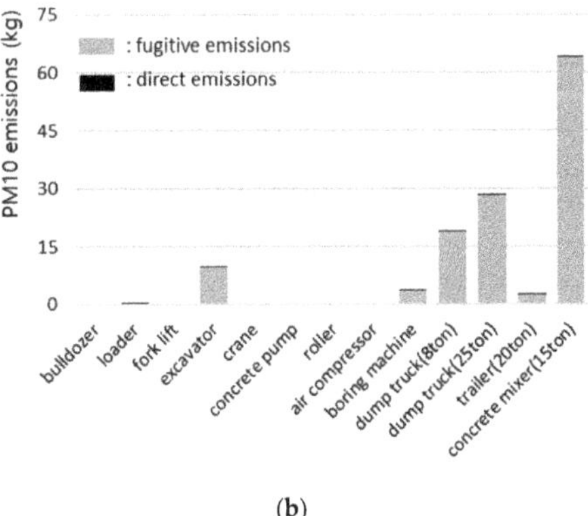

(**b**)

Figure 5. Prediction results for PM2.5 and PM10 emissions by construction equipment. (**a**) PM2.5 emissions, (**b**) PM10 emissions.

Moreover, while the direct emission factors applied in this study are reasonable since they are based on the national air pollutant emissions statistics data which are widely used in Korea, the accuracy of the fugitive emission factors needs to be verified. However, there is no proper comparison target for verification, and it is necessary to study the detailed method of fugitive emission factor development in further studies. Thus, in this study, the comparison between the fugitive emissions result of the case study and fugitive emissions calculated by Equation (1) was studied to provide basic points of fugitive emission factor development.

According to the general information of the construction site used for the case study, the annual construction area is 2564 m^2 and the annual earthwork period is 6.67 months. Since the building is a nonresidential building, the emission factors for Equation (1) are 0.0426 for PM10 and 0.00426 for PM2.5. Therefore, the total fugitive emissions calculated by Equation (1) are 7.28×10^2 kg/year for PM10 and 7.28×10^1 kg/year for PM2.5. Compared to the annual fugitive dust emissions derived in Table 7, the results of Equation (1) are approximately 5 times bigger in both PM10 and PM2.5. This gap is basically caused by the difference in the variables used in fugitive emission factor equations; in particular, the silt content and the moisture content highly affect fugitive emissions. Previously, it was found that the PM emissions increase as the silt content increases and that the PM emissions decrease as the moisture content increases [17]. According to the NIER research (NIER, 2008) for Equation (1) development, the moisture content used for fugitive emission factor calculation was 0.7% and silt content was 14.1%. However, the moisture content used in this study was 12% and the silt content was 9%, as shown in Table 2. Since the moisture content in this study is higher and the silt content is lower compared to the NIER research (NIER, 2008) the fugitive PM emissions calculated in this study were low. Thus, the moisture and silt content should be supplemented in further studies to improve the accuracy of PM emissions evaluation.

5. Discussion

This study has developed an original technology for quantitatively predicting PM emissions at construction sites. The proposed technology overcomes the limitations of the existing concentration-based PM assessment and management method. Furthermore, this study is expected to provide a guideline for investigating PM emissions at construction sites nationwide.

However, the variables applied for generating fugitive PM emission factors in this study only rely on the single study from GRI and still need to be studied to retain accuracy. Variables such as silt content and moisture content highly influence the value of fugitive PM emission factors [17]. However, methods of how the variables were selected were not considered in this study. According to the result of the case study, fugitive emissions compose the largest share of total emissions; thus, updating the fugitive emission factors by applying reasonable variables should be performed in further studies. Moreover, construction equipment considered in this study is limited to only 13 types. US EPA provides more equipment emission factors through AP-42, such as those for graders and batch plants [29]. Considering the diversity of construction events in large-scale construction sites, additional emission factors for equipment and construction work should be continuously developed in future studies.

The proposed technology is based on a limited amount of existing literature and data. Thus, further research is required to establish a more precise emission factor DB and calculate input scenarios for construction equipment. In addition, the data that are assumed in the calculation of emission factors, such as the topsoil silt content and soil moisture content, should be supplemented with geographic information.

6. Conclusions

This study aimed to develop a technology for predicting particulate matter (PM) emissions at construction sites. The primary findings of this study are as follows:

1. Thirteen types of construction equipment were selected as the main PM emission sources at construction sites. Then, an emission factor DB was established, which consisted of direct and fugitive emission factors for PM10 and PM2.5 for various types of construction equipment.
2. The PM emission activity scenarios for construction equipment were presented and used to develop a method for predicting PM emissions at construction sites.
3. A system for predicting PM emissions at construction sites was built using the emission factor DB and activity scenarios.
4. A case study was performed using the developed system, and the fugitive and direct emissions of PM2.5 and PM10 were calculated for construction equipment.
5. Moisture content and silt content values applied in fugitive factor development equations should be supplemented in further studies to improve the accuracy of PM emissions evaluation.

Author Contributions: Writing—original draft preparation, J.Y.; writing—review and editing, J.Y., H.K.; conceptualization and supervision, H.K. and S.T. All authors have read and agreed to the published version of the manuscript.

Funding: This work was supported by the Korea Agency for Infrastructure Technology Advancement (KAIA) grant funded by the Ministry of Land, Infrastructure and Transport (grant number 21CTAP-C152276-03).

Conflicts of Interest: The authors declare no conflict of interest.

References

1. Bennett, J.E.; Tamura-Wicks, H.; Parks, R.M.; Burnett, R.T.; Pope, C.A.; Bechle, M.J.; Marshall, J.D.; Danaei, G.; Ezzati, M. Particulate matter air pollution and national and county life expectancy loss in the USA: A spatiotemporal analysis. *PLoS Med.* **2019**, *16*, e1002856. [CrossRef] [PubMed]
2. Griffiths, S.D.; Chappell, P.; Entwistle, J.A.; Kelly, F.J.; Deary, M.E. A study of particulate emissions during 23 major industrial fires: Implications for human health. *Environ. Int.* **2018**, *112*, 310–323. [CrossRef]
3. Zhang, X.; Gao, S.; Fu, Q.; Han, D.; Chen, X.; Fu, S.; Huang, X.; Cheng, J. Impact of VOCs emission from iron and steel industry on regional O^3 and PM2.5 pollutions. *Environ. Sci. Pollut. Res.* **2020**, *27*, 28853–28866. [CrossRef] [PubMed]
4. Tétreault, L.-F.; Doucet, M.; Gamache, P.; Fournier, M.; Brand, A.; Kosatsky, T.; Smargiassi, A. Childhood Exposure to Ambient Air Pollutants and the Onset of Asthma: An Administrative Cohort Study in Québec. *Environ. Health Perspect.* **2016**, *124*, 1276–1282. [CrossRef] [PubMed]
5. Winquist, A.; Kirrane, E.; Klein, M.; Strickland, M.; Darrow, L.A.; Sarnat, S.E.; Gass, K.; Mulholland, J.; Russell, A.; Tolbert, P. Joint Effects of Ambient Air Pollutants on Pediatric Asthma Emergency Department Visits in Atlanta, 1998–2004. *Epidemiology* **2014**, *25*, 666–673. [CrossRef] [PubMed]
6. Rückerl, R.; Schneider, A.; Breitner, S.; Cyrys, J.; Peters, A. Health effects of particulate air pollution: A review of epidemiological evidence. *Inhal. Toxicol.* **2011**, *23*, 555–592. [CrossRef] [PubMed]
7. Pande, P.; Dey, S.; Chowdhury, S.; Choudhary, P.; Ghosh, S.; Srivastava, P.; Sengupta, B. Seasonal Transition in PM10 Exposure and Associated All-Cause Mortality Risks in India. *Environ. Sci. Technol.* **2018**, *52*, 8756–8763. [CrossRef] [PubMed]
8. Morino, Y.; Chatani, S.; Tanabe, K.; Fujitani, Y.; Morikawa, T.; Takahashi, K.; Sato, K.; Sugata, S. Contributions of Condensable Particulate Matter to Atmospheric Organic Aerosol over Japan. *Environ. Sci. Technol.* **2018**, *52*, 8456–8466. [CrossRef] [PubMed]
9. Fajersztajn, L.; Saldiva, P.; Pereira, L.A.A.; Leite, V.F.; Buehler, A.M. Short-term effects of fine particulate matter pollution on daily health events in Latin America: A systematic review and meta-analysis. *Int. J. Public Health* **2017**, *62*, 729–738. [CrossRef] [PubMed]
10. Wong, C.-M.; Vichit-Vadakan, N.; Kan, H.; Qian, Z. Public Health and Air Pollution in Asia (PAPA): A Multicity Study of Short-Term Effects of Air Pollution on Mortality. *Environ. Health Perspect.* **2008**, *116*, 1195–1202. [CrossRef] [PubMed]
11. Sun, Y.; Zhuang, G.; Tang, A.; Wang, Y.; An, Z. Chemical Characteristics of PM2.5 and PM10 in Haze-Fog Episodes in Beijing. *Environ. Sci. Technol.* **2006**, *40*, 3148–3155. [CrossRef] [PubMed]
12. Yang, X.; Lu, D.; Tan, J.; Sun, X.; Zhang, Q.; Zhang, L.; Li, Y.; Wang, W.; Liu, Q.; Jiang, G. Two-Dimensional Silicon Fingerprints Reveal Dramatic Variations in the Sources of Particulate Matter in Beijing during 2013–2017. *Environ. Sci. Technol.* **2020**, *54*, 7126–7135. [CrossRef] [PubMed]
13. Basile-Doelsch, I.; Meunier, J.D.; Parron, C. Another continental pool in the terrestrial silicon cycle. *Nature* **2005**, *433*, 399–402. [CrossRef] [PubMed]
14. Ohara, T.; Akimoto, H.; Kurokawa, J.; Horii, N.; Yamaji, K.; Yan, X.; Hayasaka, T. An Asian emission inventory of anthropogenic emission sources for the period 1980–2020. *Atmos. Chem. Phys. Discuss.* **2007**, *7*, 4419–4444. [CrossRef]

15. Zhang, Q.; Streets, D.G.; Carmichael, G.R.; He, K.B.; Huo, H.; Kannari, A.; Klimont, Z.; Park, I.S.; Reddy, S.; Fu, J.S.; et al. Asian emissions in 2006 for the NASA INTEX-B mission. *Atmos. Chem. Phys.* **2009**, *9*, 5131–5153. [CrossRef]
16. Ministry of Environment, National Air Pollutants Emission Service. Available online: http://airemiss.nier.go.kr/ (accessed on 15 January 2021).
17. Kim, H.; Tae, S.; Yang, J. Calculation Methods of Emission Factors and Emissions of Fugitive Particulate Matter in South Korean Construction Sites. *Sustainability* **2020**, *12*, 9802. [CrossRef]
18. Reff, A.; Bhave, P.V.; Simon, H.; Pace, T.G.; Pouliot, G.; Mobley, J.D.; Houyoux, M. Emissions Inventory of PM2.5 Trace Elements across the United States. *Environ. Sci. Technol.* **2009**, *43*, 5790–5796. [CrossRef] [PubMed]
19. Ministry of Environment. *Scattering Dust Management Manual*; Ministry for the Environment: Sejong, Korea, 2017.
20. National Institute of Environmental Research (NIER). *National Air Pollutant Emission Calculation Method Manual III*; NIER: Incheon, Korea, 2013.
21. The Ministry of Environment. Emergency Reduction Measures for Ultra-Fine Dust. 2019. Available online: www.cleanair.go.kr/dust/dust/dust-emergency01.do (accessed on 11 November 2021). (In Korean)
22. Ministry of Environment. Air Pollutant Emission-Cap Management for Industrial Workplace. Available online: www.keco.or.kr/en/core/climate_air2/contentsid/1947/index.do (accessed on 11 November 2021).
23. National Assembly Research Service (NARS). *Current Status of Total Air Pollution Control System at Workplaces and Improvement Plan, Legislative Policy Report*; NARS: Seoul, Korea, 2019; Volume 36.
24. Muleski, G.E.; Cowherd, C., Jr.; Kinsey, J.S. Particulate Emissions from Construction Activities. *J. Air Waste Manag. Assoc.* **2005**, *55*, 772–783. [CrossRef] [PubMed]
25. Gyeonggi Research Institute (GRI). *Seasonal Particulate Matter Management to Prevent High Pollution Events in Gyeong-gi-do*; GRI: Suwon, Korea, 2019.
26. National Institute of Environmental Research (NIER). *Air Pollutant Emission Factor: Based on 2012 Air Pollutant Emissions*; NIER: Incheon, Korea, 2015.
27. US Environmental Protection Agency. *AP-42: Compilation of Air Emissions Factors, Chapter 13.2.3: Heavy Construction Operations*; US EPA: Washington, DC, USA, 1995.
28. US Environmental Protection Agency. *AP-42: Compilation of Air Emissions Factors, Chapter 11.9: Western Surface Coal Mining*; US EPA: Washington, DC, USA, 1998.
29. US Environmental Protection Agency. *Introduction to AP 42, Volume I*, 5th ed.; US EPA: Washington, DC, USA, 1995.
30. National Institute of Environmental Research (NIER). *Improvement of Fugitive Emission Calculation Method and Development of Real-Time Measurement Method of Fugitive Emissions on Roads, Final Report*; NIER: Incheon, Korea, 2008.
31. Ministry of Land, Infrastructure and Transport. *Standard of Construction Estimate*; Ministry of Land; Infrastructure and Transport: Sejong, Korea, 2020.

 *sustainability*

Article

Effectiveness of Green Infrastructure Location Based on a Social Well-Being Index

Sanghyeon Ko [1] and Dongwoo Lee [2,*]

1   Department of Transportation Planning, Metropolitan Washington Council of Governments, Washington, DC 20002, USA; sko@mwcog.org
2   Department of Urban Policy and Administration, Incheon National University, Incheon 22012, Korea
*   Correspondence: dlee@inu.ac.kr; Tel.: +82-32-835-8738

Abstract: Urban Green Infrastructure (GI) provides promising opportunities to address today's pressing issues in cities, mainly resulting from uncurbed urbanization. GI has the potential to make significant contributions to make cities more sustainable by satisfying the growing appetite for higher standards of living as well as helping cities adapt to extreme climate events. To leverage the potentials of GI, this article aims to investigate the effectiveness of GI that can enhance social welfare benefits in the triple-bottom line of urban sustainability. First, publicly available data sets representing social demographic, climate, and built environmental elements are collected and indexed to normalize its different scales by the elements, which is termed as the "Social Well-being Index." Second, a random forest regressor was applied to identify the impacts of variables on the indexed scores by region. As a result, both the Seoul and Gyeonggi-do models found the most significant relationship with the type of GI to prevent pollutants and disasters, followed by GI types to conserve and improve the environment in Seoul and GI types to serve activity spaces in Gyeonggi-do. Furthermore, variables such as population, number of pollutants, and employment rate in Seoul were found significant and employment rate, population, and air pollution were significant in Gyeonggi-do. Finally, a scenario analysis is conducted to investigate the impacts of the overall index score with additional GI facilitation according to the model's findings. This article can provide effective strategies for implementing policies about GI by considering regional conditions. The analytical processes in this article can provide useful insights into preparing effective ecological and environmental improvement policies accordingly.

Keywords: green infrastructure; indexing; random forest; interpretation of machine learning

Citation: Ko, S.; Lee, D. Effectiveness of Green Infrastructure Location Based on a Social Well-Being Index. *Sustainability* **2021**, *13*, 9620. https://doi.org/10.3390/su13179620

Academic Editor: Luca Salvati

Received: 21 June 2021
Accepted: 23 August 2021
Published: 31 August 2021

Publisher's Note: MDPI stays neutral with regard to jurisdictional claims in published maps and institutional affiliations.

1. Introduction

As humans settle together in cities, economies of scale and efficiency have been enhanced to better provide the essentials of residents. This has made our modern cities more attractive places to live and work because they offer higher living standards compared to rural areas [1]. In general, human society essentially requires resources and energy provided by nature; however, many human behaviors in cities treat these natural resources as free and almost unlimited. For the sake of living well, human society has built many built infrastructures—also known as grey infrastructure—and it has generated other challenges we have to face in cities [2].

Cities now face numerous societal and environmental challenges, such as inequity, resource depletion, climate changes, and air quality. These recent challenges are inextricably interconnected to the challenges we have faced in the past—e.g., rapid urban population growth, pollution, limited urban infrastructure networks. To mitigate these challenges, a series of urban renaissance projects (e.g., urban retrofitting, urban regeneration) are initiated to make cities more sustainable and livable areas. For example, urban renaissance projects [3,4] aim to meet policy objectives to enhance living standards (e.g., health, social cohesion), revitalize the local economy, and help cities adapt to climate changes.

As one of the promising solutions to a series of urban renaissance projects, approaches using Green Infrastructure (GI) are being discussed and understanding of GI's potential has gained particular interests recently. Specifically, GI helps to avoid investing in a large-scale infrastructure system that requires enormous government budgets since it can often find cheaper and more durable solutions, generating various benefits. The European Commission (EU) [5], for instance, describes that the Green Space, an urban environmental indicator, is a nature-based solution that has been successfully tested in urban spaces, providing economic, social, and ecological benefits to sustainable development goals (SDGs) [6–8]. In general, GI is integrated into, or embedded within, other urban infrastructure systems to alleviate various challenges—e.g., stormwater management, carbon mitigation, energy savings in other infrastructure, air quality improvement, human health and well-being, and other social and ecological benefits. Moreover, since urban infrastructure and its consumption behaviors are intricately interconnected, GI can provide complementary solutions to existing grey infrastructure systems [2]. Therefore, GI has the potential to make significant contributions to making cities more sustainable by satisfying the growing appetite for higher standards of living as well as helping cities achieve a goal of urban sustainability.

GI holds many definitions and interpretations for different disciplines. From a planning and engineering perspective, in the realm of urban studies, it can be defined as strategically built or planned interconnected features with other infrastructure systems and environments, which can deliver a wide range of benefits to urban sustainability [9]. Several studies have concluded that whole or partially nature-based solutions, including GI and its associated elements, offer a wide range of benefits such as environmental, human health, and well-being (e.g., satisfaction) benefits in cities [1,5,10]. In terms of livability in cities, GI has various positive impacts on cities and residents [11]. The U.S. Environmental Protection Agency (EPA), for example, has also made various efforts to reduce the heat island effect in cities with vegetation, improving air quality and eliminating pollutants [12]. To put it differently, the rare integration of GI with cities has a negative impact on environmental ecosystems and the health and well-being of residents [13,14]. Based on the previous studies, it is obvious that the environmental and ecological improvements using GI can enhance residents' overall conditions. In particular, it can be viewed that the importance of GI has a wide range of effects, from the psychological satisfaction of humans to environmental health. Therefore, securing sufficient GI has a positive impact on the satisfaction of the residents of urban environments.

In addition, previous studies also found that optimal implementation of GI can be an effective strategy for challenging issues in cities—e.g., water resources depletion, air quality issues, and other extreme events associated with climate change [15]. For instance, GI can improve air quality [16] and mitigate the heat island effect in the city center [17]. Providing that GI is sufficiently installed, extreme stormwater events, heat island effects, and air pollution can be alleviated [18]. More precisely, the effectiveness of GI largely depends on the location of GI being effectively integrated into, or embedded within, other urban infrastructure systems. In particular, some studies discuss the location of green spaces and GI while considering air pollution, noise, population, temperature, and green levels [19–21]. For example, Maruani and Amit-Cohen compared open space planning models to find effective factors. They concluded that both open space factors and external factors affect planning qualities [22]. Thompson suggested new patterns of connected open space to be effective with considerations of the size, application cost, period, complexity, etc. [23] Schilling and Logan suggested a model to find right sizes of GI by community sizes [24]. Votsis and BenDor et al. [10,25] found the impact of green areas on residential selection and child-rearing were found to be highly utilizable and efficient. Kim and Tran evaluated GI's comprehensive plan performance in the United States and proposed a more comprehensive and concrete policy framework for local planners proposed for GI locations [26]. Campagnaro et al. [27] predicted the Pielou's equivalence index of Green information from Italy's Padua region, compared regional distributions according to variables, and discussed its applicability to find adequate locations. Lee and Oh [28] developed the GI location assessment system based on geographical information for Namyangju-si, Korea, and attempted the landscape ecological evaluation from both macro and micro perspectives. Zhou and Wu [29] attempted to distribute the Green Land of Hekou City, China, by matching the demand for rust and topographical characteristics. Lai et al. [30] identified the relationship between land use and spatial identification by presenting a plan to implement GIs in three areas near Cagliari, Italy.

In this circumstance, being able to analyze the effectiveness of GI is essential for municipalities and utilities to effectively meet users' needs in urban areas. In particular, GI generally occupies large lots and has difficulty changing land use once implemented [17,31], thus, disproportionately distributed GI occurs potential degradation in ecological and environmental services [32–34]. Although some previous studies include fine-grained quantitative and qualitative information about GI for analyzing the impacts of GI, it is not applicable to most cities and municipalities where cities and utilities have limited access to the fine-grained data. Therefore, in this study, we suggest an approach to help policy makers and planning experts find a more feasible solution for determining the location of GI. Basically, the aim is to investigate the effectiveness of GI in urban areas that can enhance social welfare benefits for the triple-bottom line of urban sustainability to leverage the potential of GI. Furthermore, considering the connectivity aspects of the cities, maximizing the impacts of GI by allowing it to be linked to adjacent regions under the given conditions has been determined to have a more sustainable development direction. In addition, regional differences by functional and developmental characteristics by the linkage of population expansion and urban facilities will be considered. The specific objectives of this article are to answer the following research questions:

- How does GI influence residents' satisfaction with livability?
- Which attributes of sociodemographic and regional characteristics are most sensitive for adopting GI?
- How do we maximize the model's application in the realm of urban studies?

2. Materials and Method

2.1. Materials

The overall research framework is described in Figure 1. Considering the geographical perspective of the research area, there are substantial distributional discrepancies between the two regions of Seoul and Gyeonggi-do. Thus, we made two different models to minimize technical issues in modeling—e.g., heterogeneity. To analyze the GI's influence to their residents in two different regions, we adopt a random forest (RF) regressor, which is less susceptible to scales and outliers than other machine learning (ML) approaches [35–37]. Then, the estimated ML models are interpreted by surrogate interpretation techniques. Lastly, two scenario analyses were conducted to demonstrate the applicability of our models for future policy analysis.

This study geographically focused on 25 districts in Seoul and 31 cities and counties in Gyeonggi-do, as shown in Figure 2. In terms of high population concentration and interconnected facilities, residents' responses to local GI changes were expected to be sensitive within the same living zone spectrum, thus ideal for evaluating alternatives to location green facilities presented in this work. In addition, through the relationship between better environmental conditions and population concentration, population and environmental variables are found to be highly interconnected throughout the research area, and the data should be available from the same sources to match. For reference, the detailed administrative district units of Seoul and Gyeonggi-do were marked as 'area' to represent them as unified terms in describing the data and the model setting and analysis stages.

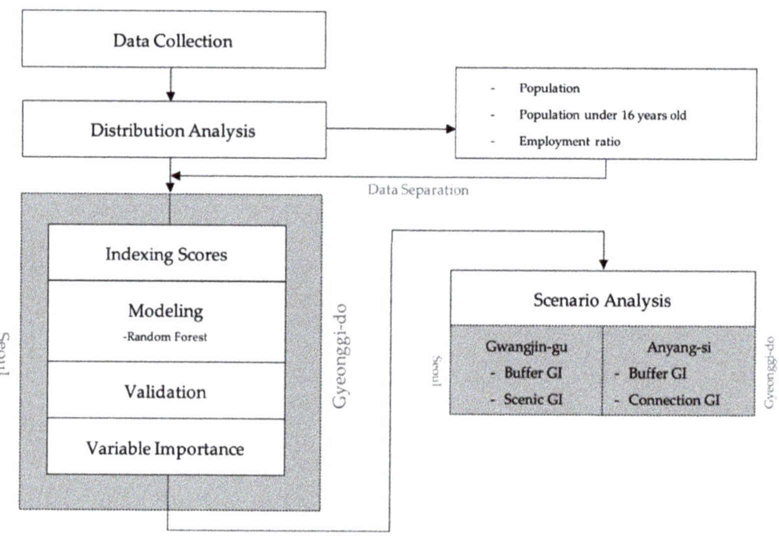

Figure 1. Research design.

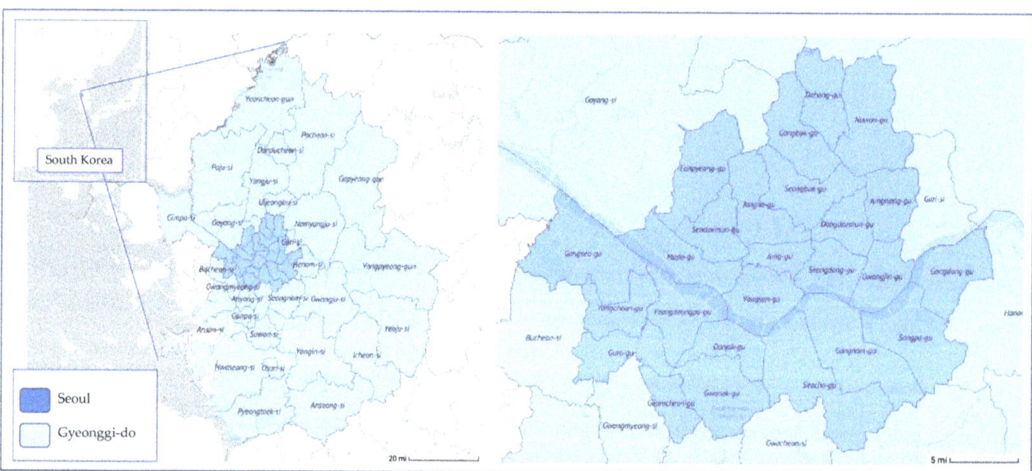

Figure 2. Geographical scope of the research area—Seoul and Gyeonggi-do in South Korea.

Studies on location selection for balanced GI expansion in cities have been mainly done through micro-level optimization approaches through multi-objective optimization by setting specific environmental factors—for example, air quality, noise, population, temperature, green scale—as variables [18,38–41]. However, the implementation of GI's facilities is characterized by having to override other projects in their location or size. It should consider a broader geographic area, including its vicinity, not just for the benefit of the region, which is the reason to provide information to consider the GI's location at a higher level of the planning process to plan GI location preferentially [22]. In addition, various and ever-changing environmental factors should be included as much as possible to be analyzed in more detail.

Among the publicly available sources, data of demographics, GI, climate, and built environmental characteristics were collected, which can be used to check the environmental and ecological conditions of each region and area. To capture the concentration of the population in the city and its relationship or preference with climate and environmental variables in consideration of the population per unit area in each region, the area's total numbers of the population, population under 16 years of age, and the employment ratio was considered. The annual data of GI, road rate, and the number of pollutants producing facilities were utilized as they were found to have an ecological impact on the urban environment. Green areas and the number of air pollution and water pollution emission facilities per area unit were considered. The Buffer GI is created to prevent pollution and natural disasters. The Connection GI serves as a walking space for urban residents by organically connecting parks, rivers, and mountainous areas. The Scenic GI is installed for environmental conservation and improvement. The three types of green areas are divided into Buffer GI, Connected GI, and Scenic GI by laws and regulations of the Ministry of Land, Infrastructure and Transport, the Ministry of Environment, and local governments. Since the scale of installation for each type of green area is also different, all three types of GIs were considered as variables differently in this study.

The climate variables were taken into account by the average monthly temperature (in Celsius), wind speed (meter/second), and rainfall (millimeter). The most stable air pollutants were collected using PM10 (Particulate Matter, finer than 10 micrometers) average monthly data (microgram/m^2). The number of air and water pollution emission facilities for each area were collected. Considering the conventional time zone of the entire data, the time range was set from January 2013 to December 2019. The variables and their descriptive statistics are shown in Table 1.

Table 1. Descriptive statistics of independent variables.

Variables	Description	Time Range	Mean	S.D.	Min.	Max.
	Demographic characteristics					
Pop	Total population	Year	404,873.60	252,774.03	43,824	1,203,285
Pop_16	Number of the population age 16 or under	Year	61,644.73	43,861.39	5020	213,391
Emp	Employment ratio (employed/total population)	Year	59.31	2.37	49.6	66.5
	Green Infrastructure characteristics					
Buffer_gi	Buffer Green Infrastructure area (m^2)	Year	573,978.77	815,943.75	0	4,504,737
Scenic_gi	Scenic Green Infrastructure area (m^2)	Year	494,455.95	2,672,348.84	0	34,908,592
Connection_gi	Connection Green Infrastructure area (m^2)	Year	93,391.83	427,795.86	0	3,939,590
	Built environment characteristics					
Water_pol_fac	Number of water pollution facilities	Year	320.99	409.81	19	2775
Air_pol_fac	Number of air pollution facilities	Year	314.86	530.14	2	3231
Road	Road coverage (km^2)	Year	11.65	10.21	0.11	28.83
	Climate characteristics					
temp	Temperature (°C)	Month	12.51	10.12	−8.5	29.3
wind	Wind speed (m/s)	Month	1.54	0.58	0	6.8
precipitation	Precipitation (mm/day)	Month	85.07	108.34	0	1076.50
PM10	Particulate Matters (<10 micrometers, micrograms/m^2)	Month	47.39	15.81	6	124

Although there is no significant difference between weather and air pollution as Gyeonggi-do shares the same geographical characteristics with the boundary of the form surrounding Seoul as shown in Figure 1, it should be considered that there is a difference in the development process of the two regions [40].

To compare heterogeneous demographic characteristics between Seoul and Gyeonggi-do, bivariate probabilistic density for the two regions was estimated. Joint probability density estimation offers a means to study the joint probability density distribution between two variables (i.e., bivariate distribution) in a nonparametric way. This is significant because the joint probability density distribution between two variables captures the relationships between the two variables. In other words, it shows the probability by which each of the two variables falls in any certain intervals or discrete sets of values. For instance, suppose that X and Y are continuous variables, and X takes values in $[a, b]$ and Y takes values in $[c, d]$. Then the pair of X and Y takes the product $[a, b] \times [c, d]$. In this case, we can express the joint probability distribution of two variables, X and Y, as:

$$\int_c^d \int_a^b f(x, y)dx\, dy = 1 \tag{1}$$

In addition to studying the joint probability density distribution of two variables, we estimate a univariate probability density function for each single variable. Specifically, the univariate probability density function of a single variable, X, can be estimated using a kernel density estimation method. For instance, the estimated function $(\hat{f})$ for the variable, $X = \{x_1, x_2, \cdots x_n\}$, is derived from the kernel density estimator as defined below:

$$\hat{f}_{(h)}(x) = \frac{1}{n}\sum_{i=1}^{n}K_h(x - x_i) = \frac{1}{nh}\sum_{i=1}^{n}K(\frac{x - x_i}{h}) \tag{2}$$

where K is the kernel or weight function (it should be a non-negative function that integrates into one), and h is a smoothing parameter that acts as a bandwidth. For the kernel (K) and the smoothing parameter (h), we apply the standard normal kernel and auto-selection algorithm derived from the asymptotic mean integrated squared error (AMISE). In Figure 3, the bivariate probability distribution of demographic variables for the two regions was estimated.

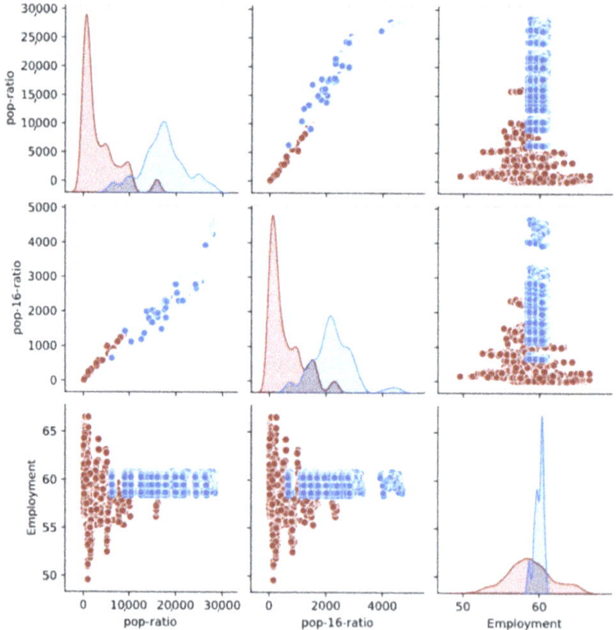

Figure 3. Demographic bivariate distribution for Seoul and Gyeonggi-do (Seoul in blue and Gyeonggi-do in red).

The total population and the number of people under the age of 16 per unit area were clearly different in Seoul and Gyeonggi-do, and the comparison of employment rates showed that Seoul was concentrated in the 60% range, while Gyeonggi-do was widely spread around the late 50%. In terms of road rate, it was also confirmed that the road rate per unit area in Seoul was much higher than that of Gyeonggi-do. The relationship between the variables in Seoul and Gyeonggi-do were distributed and noticeably distinguishable from the difference between the characteristics of the two areas. Therefore, since it is reasonably concluded to consider the location of GI by the characteristics between the two regions, separate models were prepared.

2.2. Model Method

The selected and collected measurement data have different characteristics to compare as they are and they have difficulty considering weights. The alternative approach to resolve this conflict is the indexing method for measuring Better Life Index in the Organisation for Economic Co-operation and Development (OECD) [41] and Seoul, Korea [42]. In this study, this indexing method was applied to normalize the scale in the relevant variables and to utilize the sum of the entire scale as an index for analysis. It was named the Social Well-Being Index. Through the RF Regression, the importance of variables affecting index were examined by specifying the sum of index obtained by month and region as dependent variables. Seoul and Gyeonggi-do, which were established as research areas, were composed of separate models because the distribution of population and geographical characteristics were not the same. The model was constructed using data from 2013 to 2017, and it was validated using data from 2018 and 2019. Based on these estimated models and the importance of identified variables, a scenario analysis was conducted varying regionally important GI in Seoul and Gyeonggi-do to determine how much improvement is made in the Social Well-Being index in each region. Although a direct model for GI is not suggested, it is considered a more effective method for sustainable planning because it measures the expected effects of the expansion of GI in situations where both climate and environmental variables are considered.

2.2.1. Social Well-Being Index

Indexing is the process of leveling the range of analysis and identifiable indicators among the variables that affect climate and environmental condition in this research context. Since it is impossible to directly compare the original values of the data to each other because all the variables for the observed climate and environmental indicators were measured at different levels and ranges, the level and scope of the measurement are identified for correct comparison. In this process, the method of equalizing the values of the data was applied to the Min-Max method used by the OECD and Seoul, Korea, and variables will have index scores based on the following Equation (3):

$$Positive\ Var.\ Index = \frac{x_i - min(x)}{max(x) - min(x)} \tag{3}$$

where x represents the input records ($x = x_1, x_2, \cdots, x_n$).

As a result of this Equation (3), the index value is located between 0 and 1 regardless of the unit and distribution of the corresponding variable origin values enables comparison between variables. However, if the characteristics of the variables that make up the data have a negative meaning, such as the values of the pollutant facilities, air quality, and road rates, the corresponding indicators were used as minus the original equivalence value as shown in the Equation (4):

$$Negative\ Var.\ Index = 1 - \frac{x_i - min(x)}{max(x) - min(x)} \tag{4}$$

The index scores for all variables were obtained, and Equation (5) represents the

$$Social\ Well - Being\ Index = \sum Positive\ Var.\ Index_i + \sum Negative\ Var.\ Index_j \tag{5}$$

This process cancels the scale difference of the variables, and sets the sum of them, the Social Well-Being Index score, as the model's dependent variable. Through this, each variable was exploited to determine the impact and relationship on the overall index.

As aforementioned, we used the sociodemographic and geographical characteristics of Seoul and Gyeonggi-do and the correlation among variables for two regions presented in Figure 4. The population in Seoul is highly correlated with a population under 16, showing a correlation at 0.92, but it is seen as a normal relationship when viewed at the rate of population size. Weather-related factors including precipitation and temperature of PM10, where inverse index scale is proposed, were all are found to be in the correct relationship. The relationship between air pollution source facilities and water pollution source facilities is also seen as a proper relationship because one facility often releases air and water pollutants simultaneously. Similar correlation patterns in population, weather, and pollutant facilities can be found in Gyeonggi-do, just as in Seoul. However, the difference from the Seoul area is that the road rate shows a high inverse correlation compared to the number of people. Considering that the road rate is inversely proportional, the road rate is higher in places with a large total population and a large population under the age of 16 and is more correlated than in Seoul. Unlike Seoul, Gyeonggi-do has developed arterials before its residential area, and this is also a proper relationship because public transportation networks are also not closely connected compared to Seoul. The two regional differences are also distinct in the correlation between the population, land use, and variables in Seoul and Gyeonggi-do. Therefore, we can see that developing a model for each of the two regions is the correct analysis approach.

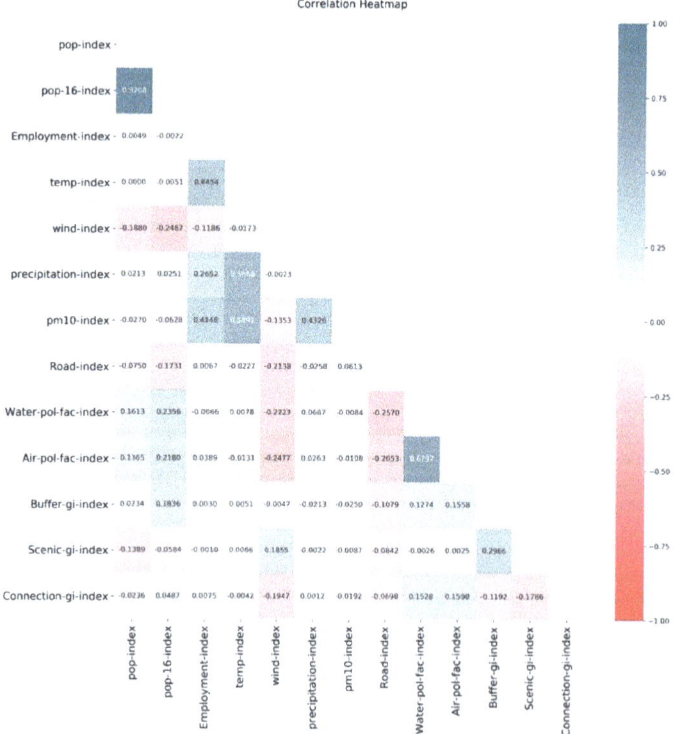

Figure 4. *Cont.*

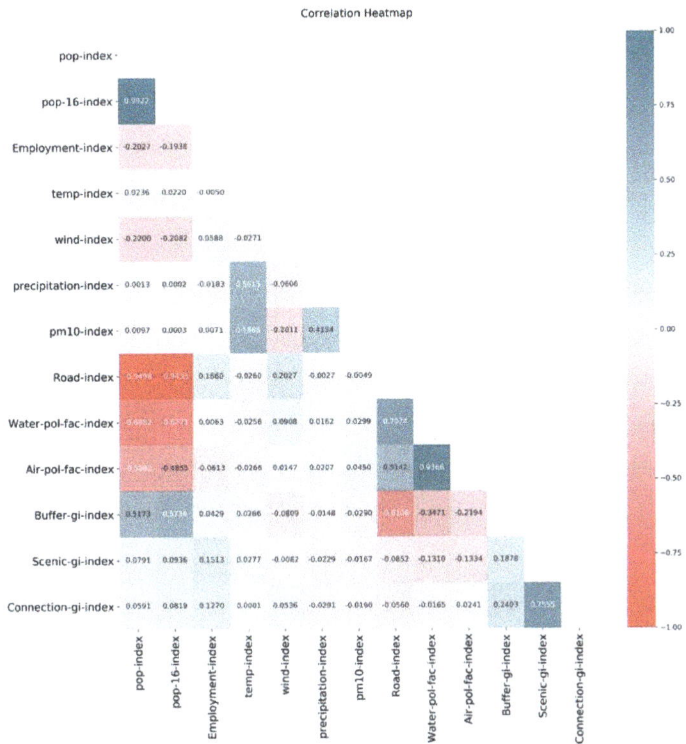

Figure 4. Correlation between variables in two regions (Relationship of positive in blue and negative in red; above: Seoul; below: Gyeonggi-do).

2.2.2. Random Forest Regressor: Rule-Based Bootstrap Aggregation

Tree- and rule-based models, being applied to both classification and regression problems, are nonparametric modeling methods (also known as machine learning, ML) [43,44]. This nonparametric additive model learns the nonlinear relationships between a set of predictors and corresponding targets based on specific splitting rules without predetermined assumptions—e.g., linearity.

In particular, a regression tree (also known as a decision tree regressor), for example, is constructure by recursively partitioning the predictor space (i.e., samples) into several homogeneous groups (also known as tree nodes), which ends up with terminal nodes.

The goal of regression trees is to find non-overlapping partitions (R_j) that minimize the overall sum squared errors, given by:

$$SSE = \sum_{j=1}^{J} \sum_{i \in R_j} (y_i - \hat{y}_{R_j}) \tag{6}$$

where $\hat{y}_{R_j}$ is the average response of the training observations within the jth partition. A single tree model locally learns patterns, i.e., local learners, and then aggregates local learners as a hierarchical tree structure (i.e., additive learning).

Thanks to the logic of their estimation, i.e., the set of splitting rules used to partition the predictor space are provided with a hierarchical structure (i.e., a tree), tree-based models have relative competitiveness with the other modeling approaches for various reasons. First, tree models built on an additive and hierarchical learning algorithm have gained popularity because of their ability of recognizing the non-linear patterns (e.g., non-linear relationships) of input features, while providing a higher degree of interpretability [44]. Furthermore, tree models can effectively handle various types of variables in nature (e.g., continuous, categorical, sporadically missing values) but also estimate models without having enough observations compared to other ML approaches—e.g., artificial neural networks [35,44].

Nonetheless, models learned by single trees or rules (e.g., regression tree) have two acknowledged weaknesses compared to other supervised learning approaches: (1) instability of model estimation and (2) relatively low predictive performance compared to other ML models. Specifically, the hierarchy of variables seen in a regression tree can vary each time a tree is estimated. In general, to train a regression tree, a certain portion of the data is commonly used, and the remaining portion is kept to test the accuracy of the tree, which is known as the "hold out" method. Therefore, if a different portion of the data is used to generate another tree, the hierarchy of variables changes. As a result, fitting one single tree is not enough to validate a hierarchy of variables. In addition, the latter is due to the fact that local fitting processes (i.e., recursively partitioning sub-regions) can increase the variance of the overall tree structure, which causes higher prediction errors than other ML models—i.e., bias-variance trade-off [45].

To alleviate these weaknesses of single tree models, bagging or bootstrap aggregation techniques (also known as ensemble methods) can be applied to balance the overall bias and variable of models [35,45,46]. RF regressor is basically an aggregate of bootstrapped trees (i.e., tree estimators) with decorrelating processes: (1) each bootstrapped tree is based on a random sub-sample of the given observations, and (2) each partition within each bootstrapped tree is split by a random subset of predictors. Thanks to this decorrelating process, RF can provide an additional improvement over simple bagging algorithms by reducing the correlation among the collection of bagged trees. RF regressor predicts y given x_i as follows:

$$y\left(x_i\right) = \hat{f}_{RF}^{N}(x_i) = \frac{1}{N} \sum_{b=1}^{N} T_b(x_i) \tag{7}$$

where x_i is a vector of an independent variable, $T_b(x_i)$ represents a single regression tree grown by bootstrapped samples including a subset of variables, and N represents the total number of trees.

2.2.3. Interpretation of ML Models

As mentioned above, RF tends to show high predictive performance thanks to its machine-based repetitive computation (also known as a bagging process) [47]. Nonetheless, high predictive performance (e.g., prediction accuracy) does not provide enough information to investigate the impact of policies and associated users' behaviors, especially in the realm of urban studies. To enhance the applicability of ML, urban analytics with ML also require an understanding of how model makes a certain decision based on the given data samples.

In predictive modeling, the contribution of predictor variables can be varied. In general, a few variables have substantial impacts on the result. To obtain useful information about the relative contribution of each variable, the reduction in bias and variance is measured, which is referred to as the "Variable Importance" (VI) (also known as the feature importance).

The sum of squared residuals (SSR) can be defined as:

$$SSR = \sum_{c \in leaves\ (T)} \sum_{i \in c} (y_i - y_c)^2 \tag{8}$$

where i is an observation on leaf c, and y_i is the predicted value of the dependent variable of observation i. y_c is the actual value of the dependent variable of leaf c. The changes in SSR among the variables indicates the VI of x_j and it can be calculated as follows:

$$VI_d\left(x_j\right) = \Delta_d = SSR_d - \sum_i SSR_i^d \tag{9}$$

where d denotes a node, i is the branch of node d, SSR_d is a terminal node (i.e., leaf node), and SSR_i^d is an internal node. For the entire tree model, the VI score for each variable can be expressed as follows, where D is the total number of nodes:

$$VI\left(x_j\right) = \frac{\sum_{d=1}^{D} VI_d\left(x_j\right)}{n\_nodes} \tag{10}$$

In general, the VI scores are expressed with the standardized values—i.e., ranging from 0 to 1. For instance, the VI score becomes zero (i.e., $VI_d\left(x_j\right) = 0$) when a variable has no "contribution". This interpretation can provide useful insights into planning, regulatory factors, and policy impacts in the realm of urban studies. The SHAP package was utilized to visualize VI scores more readable [48].

3. Model Results

Based on the data from 2013 to 2017, the Social Well-Being Indices of Seoul and Gyeonggi-do were estimated through the RF Regressor. A total of 13 variables were used to build the model, including Buffer GI, Connection GI, Scenic GI, water and air pollutants facilities, road rates, wind speed, temperature, precipitation, air quality, total population, the population under 16 years of age, and the employment rate. As a result of validating the model using data from 2018 and 2019, a comparison of the estimated values of the Social Well-Being Index for each region and the model's prediction values are shown in Figure 5.

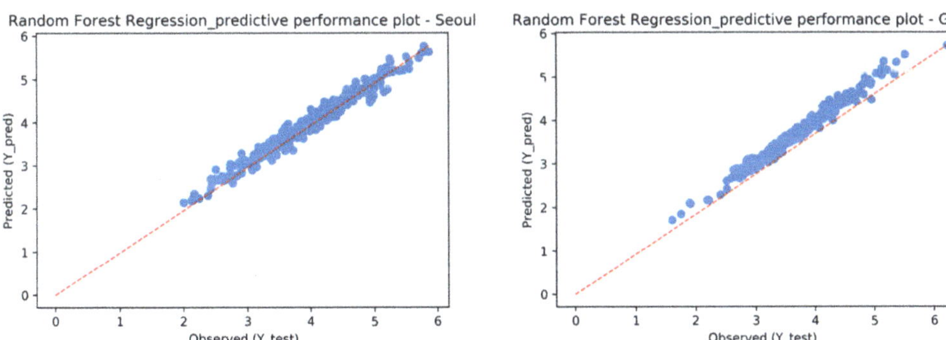

Figure 5. RF model performance plot (**Left**—Seoul, **Right**—Gyeonggi-do).

Figure 5 presents plots between the estimated Social Well-Being Index value and the modeled value of 2018 and 2019, with an R-squared of 0.97 for the Seoul model and 0.98 for the Gyeonggi-do model. The explained variance score for the Seoul model is 0.96 and is 0.98 for Gyeonggi-do model.

3.1. Relative Importance of Variable

In addition to the model's high predictive accuracy, the VI is prepared to identify variables showing significant impacts on each model. As shown in Figure 6, the importance of the Seoul model was in the order of temperature, population under 16, water pollution source, employment, Buffer GI, precipitation, and Scenic GI, Gyeonggi-do model was in the order of temperature, Buffer GI, Connection GI, employment, population, PM10, and precipitation. It is interesting to see that the temperature in both regions has the most significant impact on the overall model prediction. Regarding GI, both the Seoul and Gyeonggi models were found to significantly impact the Buffer GI compared to other GI models. Due to the characteristics of Buffer GI, it was judged that the more it was installed to distinguish it from the contaminated or pollution emitting facility, the more it affected the entire index. Scenic GI seems to have a positive impact in the Seoul model because development has already been completed, and the concentration of residents in urban areas is high. However, the competition model showed higher importance of Connection GI connecting to green areas outside the city center, indicating that the shape of GI affecting the index was also different.

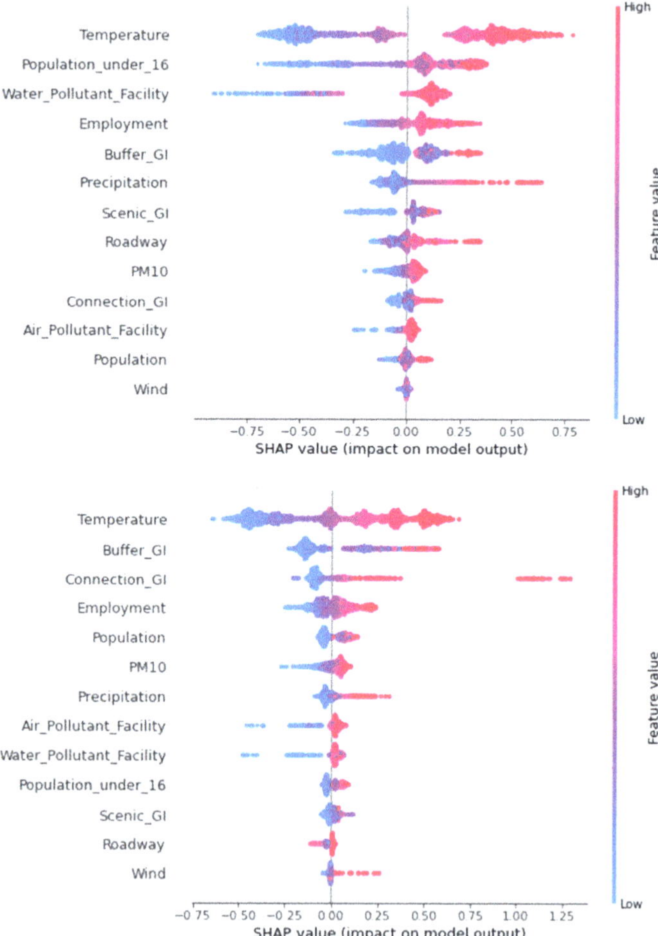

Figure 6. Importance of variables (**Upper**—Seoul, **Lower**—Gyeonggi-do).

3.2. Green Infrastructure Index Distribution

A visual representation is prepared in Figure 7 with the total index by region and the index values by the shape of GI to identify differences in regional distributions.

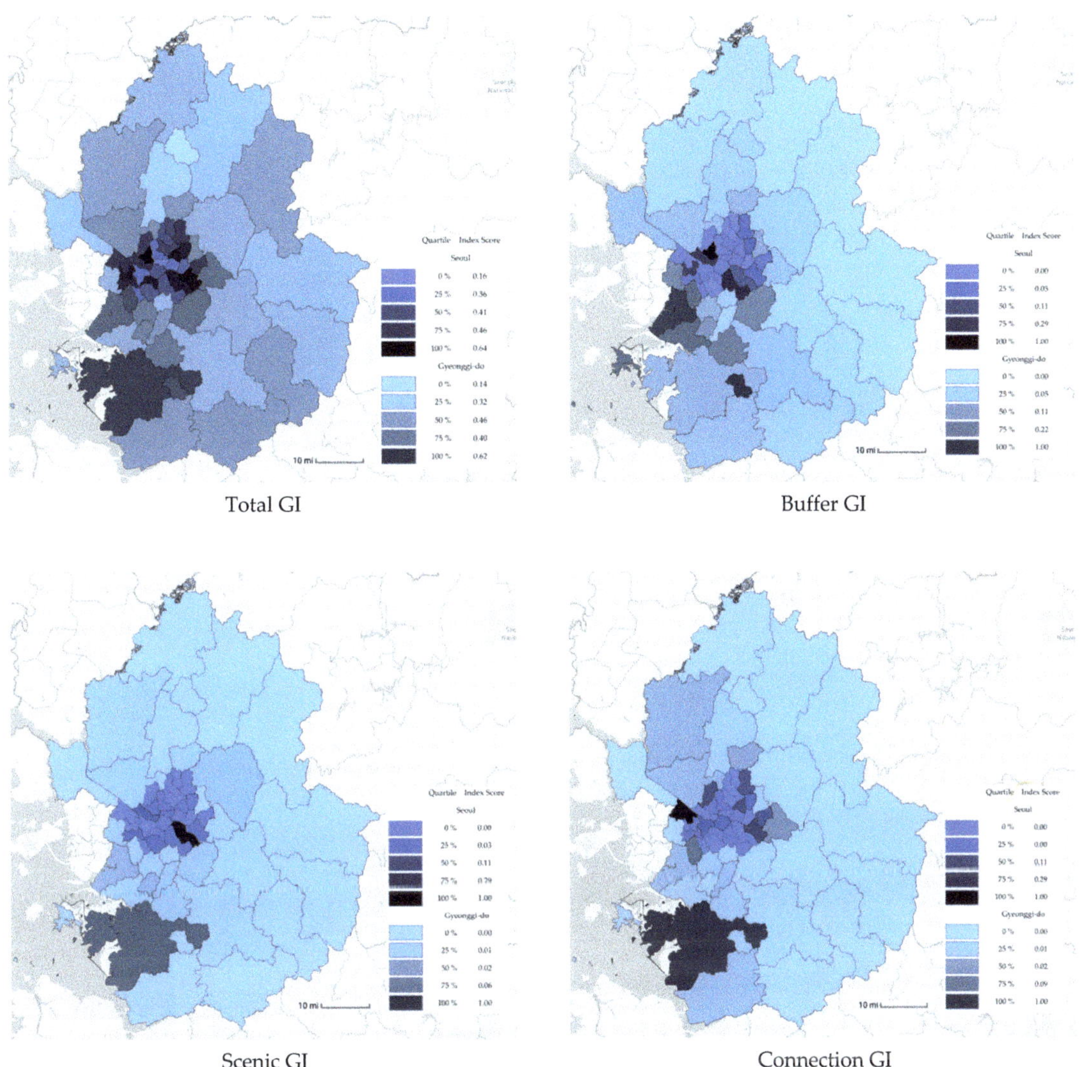

Figure 7. Social Well-being Index score distribution.

Comparing the Social Well-Being Index scores, Seoul generally showed higher scores in most regions. On the other hand, Gyeonggi-do showed higher index scores in the south than in the north. It is easy to see that the index value of each GI varies from region to region. For example, the buffer GI was found installed in areas that include airports and major highways in Seoul. In contrast, Gyeonggi-do was found to represent high scores in areas near highways and industrial complexes entering Seoul. Connection GI obtained a high index score in some areas—i.e., the connection of green areas among parks. However, Scenic GI is in one of Seoul's highest areas where a particularly high availability of space has been confirmed. In addition, various types of GI are sporadically distributed in Seoul. Still, in Gyeonggi-do, the distribution of GI is located adjacent to Seoul and more in the southern region than in the northern part.

4. Scenario Analysis and Discussion

4.1. Scenario Analysis: Potential Impacts of GI on Social Well-Being Index

As mentioned above, a model was designed to measure the Social Well-Being Index by using publicly accessible information, representing demographic, green infrastructure, climate, and built-environment characteristics. The models were verified by a 10-fold cross-validation process. Based on the estimated models, a scenario analysis was designed to further analyze the impacts of GI for the Social Well-Being index score. Specifically, the impacts of the Social Well-Being Index score resulted from the changes in the GI ratio were measured by selecting a specific area in each region. Climate conditions, including temperature, wind speed, precipitation, and demographic conditions, are assumed to be fixed. The scenario is established considering only the magnitude change of the GI types, the main object of this study. Too many combinations are expected for scenario analysis across the region, and the performance of model effects is not easy to observe. Therefore, each area with a low GI ratio and a high population ratio in Seoul and Gyeonggi-do was selected. The effect of the Social Well-Being Index on more residents when the area of GI is expanded by region is examined. As a result of checking the most recent record in the data, Gwangjin-gu in Seoul and Anyang-si in Gyeonggi-do were selected based on data from December 2019. Considering the differences in regional GI importance through the VI of RF regressor as shown in Figure 6, each scenario is prepared by adjusting the ratio of Buffer and Scenic GIs for Gwangjin-gu and adjusting the Buffer ratio Connection GIs for Anyang-si. The selected area's GIs are set as the default outcome—Base Case—and three different scenarios are prepared: the mean of the region's unit area, 150% of the mean, and the highest unit area. Therefore, Gwangjin-gu, Seoul, has set a scenario that applies Buffer GI and Scenic GI to Seoul's mean, 150 percent of the mean, and the highest ratio in Seoul. Anyang-si, Gyeonggi-do, confirmed the change in Social Well-Being Index scores by setting up and applying Buffer GI and Connection GI three scenarios, which apply the mean, 150% of the mean, and the highest ratio in Gyeonggi-do.

The overall scenario design and analysis results can be found in Table 2 and the Social Well-Being Index also improved as the proportion of GI increases. In the case of Gwangjin-gu, Seoul, the improvement of the Social Well-Being Index was confirmed in all three scenarios because the area ratios of Buffer and Scenic GI were observed to be lower than the average of Seoul and enough spaces for the GI was available for its Social Well-being index score improvements. Interestingly, due to the relatively low GI ratio compared to other districts in Seoul, the Index value represented the highest rank in the entire city as the GI area ratio increases. Anyang-si, Gyeonggi-do, can see that if Buffer and Connection GIs are adjusted to the mean of Gyeonggi-do, the ratio of Buffer GI was lower than the current one, and the ratio of Connection GI was higher than the present condition, resulting in no change due to the trade-off result in the overall Index values. Instead, when the GI area ratio is 150%, and the maximum of the mean of Gyeonggi-do, the rise of Social Well-Being Index rank resulted in being proportional to the increase in GI area ratio. In conclusion, when each area in Seoul and Gyeonggi-do was selected to analyze the scenario cases and scrutinize the results, the changes and predictions in the model do not deviate significantly from the previous verification results and are confirmed to be reliable. In addition, rather than randomly increasing the proportion of GI, the Social Well-Being Index can significantly promote the efficiency of GI positioning and sizing and maximize its effectiveness. Residents' livability can be increased in the process of presenting a location to install GI that can achieve the appropriate size and a high efficiency, and it will be a valuable reference in the policy-making process. Finally, upon this process being undertaken appropriately, the burden of post-GI location changes will be alleviated, enabling more active policy implementation and sustaining policy direction to maintain viability through proper form of GI installation in line with long-term land use development.

Table 2. Scenario variables and Social Well-being Index ranks.

Analysis Cases		GI Type	Gwangjin-gu (Seoul)		Anyang-si (Gyeonggi-do)	
			Ratio	Index Rank	Ratio	Index Rank
Base case (December 2019)		Buffer	0.099%		0.551%	
		Scenic	0.011%	10	0.219%	28
		Connection	0.000%		0.043%	
Scenario cases	Mean ratio	Buffer	0.506%		0.520%	
		Scenic	0.151%	9	0.219%	28
		Connection	0.000%		0.055%	
	150% of the mean ratio	Buffer	0.759%		0.780%	
		Scenic	0.226%	8	0.219%	26
		Connection	0.000%		0.083%	
	Maximum ratio	Buffer	2.467%		1.969%	
		Scenic	1.062%	1	0.219%	6
		Connection	0.000%		0.506%	

4.2. Discussion

This article aims to investigate the impacts of GI and the relevant factors affecting life satisfaction in cities by adopting ML approaches (i.e., random forest regressor) with surrogate interpretation. Specifically, the model is designed to investigate an index score of variables for various characteristics and scales, suggesting the analytical potential to represent the overall well-being state of the city and the availability of GI location choices in a sustainable urban space. Existing GI studies mainly considered geographic and topographic characteristics as key variables [26–30]. In this study, however, publicly accessible descriptive variables—i.e., demographic, climate, environment, and form-specific GI variables—were considered together to enable the overall efficiency improvement of urban elements, including residents. This revealed that the variables utilized in the model and Social Well-Being Index have different regional relationships and confirmed that the different GI types are found important in the models. Further scenario analysis allowed us to verify the effectiveness of improving Social Well-Being Index efficiently when expanding GIs that we considered to be regionally important. In this context, it is possible to choose a location for GI that increases equity with the surrounding regions and indicated that it was effective in establishing sustainable development plans for planners.

In addition, the findings from this article can help planners consider and evaluate efficient and proper location of the GIs in the Social Well-being perspectives. The results of the study suggested that effective location selection of GIs is also effective across urban areas. Recently, research on the effectiveness and location of GIs has been highly interested in, and efforts have been made to apply them to planning processes through various approaches. In addition, as various data become available, models that take many variables into account are presented and the application feasibility is emphasized through case examples. However, we believe that the model that does not consider residents and their circumstances, the largest and essential beneficiaries of GI, deviates from the core of sustainable development in urban areas where new natural GI cannot be found. Selecting a location in consideration of the type of household composition of residents, employment conditions, road accessibility, and climate and environmental factors in their lives can secure the sustainability of existing and new GI in cities. Residents' livability can be increased in the process of presenting a location to install GI that can achieve appropriate size and high efficiency, and it will be a valuable reference in the policy-making process. Finally, upon this process being undertaken appropriately, the burden of post-GI location changes will be alleviated, enabling more active policy implementation and sustaining policy direction to maintain viability through proper form of GI installation in line with long-term land use development.

Lastly, predictive models have been integral for urban infrastructure planning as well as decision-making process in urban contexts to investigate urban behaviors and phenomena. Nonetheless, researchers generally face technical challenges when analyzing real-world urban problems, such as the availability of data and modeling techniques. Although we are living in the era of Big Data, most cities and utilities generally do not have access to fine-grained information about the characteristics of urban areas. In this context, this article can provide useful insights into constructing acceptable models using publicly accessible descriptive variables for understanding the relationship between users' satisfaction and GI systems in an urban context.

5. Conclusions

In this study, data on climate and environmental factors as well as demographics and pollutant facilities were quantified through indexing before the RF regressor was used to identify GI's ecological and environmental factors to present more objective evaluation criteria in selecting GI locations. As a result of the models for Seoul and Gyeonggi-do, by GI types, both models have the largest impact by Buffer GI, followed by Scenic GI in Seoul and Connection GI in Gyeonggi-do. In addition, variables such as population, number of pollutants, and employment rate in Seoul were found significant, and Gyeonggi-do had a big impact on employment rate, population, and air pollution. As a result, the improvement of the overall index score was identified, indicating a noticeable improvement in the Ecological Index (see Figure 7).

Based on the data and analysis results, it was possible to identify efficient GI installation conditions by region and present criteria to maximize GI installation effectiveness based on climate and environmental improvement policies. It is also noteworthy that this research has identified the possibility of a local Social Well-Being Index to analyze utilizing existing measurements and collection data, rather than requiring additional and more resources to check the weights of variables. The three issues promised in the introduction of the study were also identified: (1) how GI affects residents' satisfaction; (2) what variables are most sensitive to GI installation; and (3) how to maximize the actual application of the model. The index score was calculated to measure social well-being standards and the model predicted their major impacts. The GI variables were found to be significant among all and the significance was different by Seoul and Gyeonggi-do. Sensitive variables associated with GI installation were demographic variables, and additional needs were identified according to the existing GI installation level. Scenario analysis also highlighted that the practical applicability of urban research and concluded the ML-based model showed impressively high predictive accuracy and identified the region's characteristics by establishing a distinguished model of Seoul and Gyeonggi-do from the inconsistent distribution characteristics of regional data. It is encouraging that such built models are also sensitive in real-case scenario analysis, confirming the potential to help decision makers make policy decisions. However, there is still a possibility of expansion in collecting extensive data considering the diverse variables of urban context including ecologic and environmental variables, time range, and transferability of the model prepared for the Social Well-Being Index. Furthermore, it is expected that more efficient GI location exploration will be possible through a comprehensive model construction that considers these variables. Finally, detailed standards for urban development and facility preparation can be provided if a more detailed model is estimated according to the applied classification rather than the type of GI.

Author Contributions: Conceptualization, methodology, validation, formal analysis, writing of the original draft preparation, and writing—review and editing, S.K.; funding acquisition, methodology, data curation, investigation, writing—review and editing, validation and supervision, D.L. Both authors have read and agreed to the published version of the manuscript.

Funding: This research received no external funding.

Data Availability Statement: Anyone with minimal experience in statistical modeling should be able to use the ML models used in this study easily by using this Python package or most other libraries available in Python and in other computer languages (e.g., R). The Python codes developed for this study are available from the corresponding author upon request.

Acknowledgments: The views and opinions of authors expressed herein do not state or reflect those of the Metropolitan Washington Council of Governments (MWCOG)/Transportation Planning Board (TPB.)

Conflicts of Interest: The authors declare no conflict of interest.

References

1. Allen Klaiber, H.; Phaneuf, D.J. Valuing Open Space in a Residential Sorting Model of the Twin Cities. *J. Environ. Econ. Manag.* **2010**, *60*, 57–77. [CrossRef]
2. Pauleit, S.; Ambrose-Oji, B.; Andersson, E.; Anton, B.; Buijs, A.; Haase, D.; Elands, B.; Hansen, R.; Kowarik, I.; Kronenberg, J.; et al. Advancing Urban Green Infrastructure in Europe: Outcomes and Reflections from the GREEN SURGE Project. *Urban For. Urban Green.* **2019**, *40*, 4–16. [CrossRef]
3. Department for Transportation, Local Government and the Regions. *Towards an Urban Renaissance*; Taylor & Francis: Abingdon, UK, 1999.
4. Imrie, R.; Raco, M. *Urban Renaissance—New Labour, Community, and Urban Policy*; The Policy Press: Bristol, UK, 2003.
5. European Commission; Expert Group on the Urban Environment. *European Sustainable Cities*; Nuclear Safety and Civil Protection: Brussels, Belgium, 1996.
6. United Nations. *The Sustainable Development Goals Report*; The Department of Economic and Social Affairs: New York, NY, USA, 2020.
7. Department of Economics and Social Affairs. *The Future Is Now—Science for Achieving Sustainable Development*; United Nations: New York, NY, USA, 2019.
8. Derrible, S. *Urban Engineering for Sustainability*; The MIT Press: Cambridge, MA, USA, 2019; ISBN 978-0-262-04344-1.
9. City of Portland. *Portland's Green Infrastructure: Quantifying the Health, Energy and Community Livability Benefits*; City of Portland: Portland, WA, USA, 2010; p. 101.
10. Votsis, A. Planning for Green Infrastructure: The Spatial Effects of Parks, Forests, and Fields on Helsinki's Apartment Prices. *Ecol. Econ.* **2017**, *132*, 279–289. [CrossRef]
11. Botkin, D.B.; Beveridge, C.E. Cities as Environment. *Urban Ecosyst.* **1997**, *1*, 3–19. [CrossRef]
12. Sandström, U.G. Green Infrastructure Planning in Urban Sweden. *Plan. Pract. Res.* **2002**, *17*, 373–385. [CrossRef]
13. United States Environmental Protection Agency (US EPA). *Benefits of Green Infrastructure*; US EPA: Washington, DC, USA, 2021.
14. Tzoulas, K.; Korpela, K.; Venn, S.; Yli-Pelkonen, V.; Kaźmierczak, A.; Niemela, J.; James, P. Promoting Ecosystem and Human Health in Urban Areas Using Green Infrastructure: A Literature Review. *Landsc. Urban Plan.* **2007**, *81*, 167–178. [CrossRef]
15. Wu, J.; Plantinga, A.J. The Influence of Public Open Space on Urban Spatial Structure. *J. Environ. Econ. Manag.* **2003**, *46*, 288–309. [CrossRef]
16. Mell, I.C. Can Green Infrastructure Promote Urban Sustainability? *Proc. Inst. Civ. Eng. Eng. Sustain.* **2009**, *162*, 23–34. [CrossRef]
17. Sung, H.-C.; Hwang, S. A Preliminary Study on Assessment of Urban Parks and Green Zones of Ecological Attributes and Responsiveness to Climate Change. *J. Korea Soc. Environ. Restor. Technol.* **2013**, *16*, 107–117. [CrossRef]
18. Zhang, Y.; Murray, A.T.; Turner, B.L. Optimizing Green Space Locations to Reduce Daytime and Nighttime Urban Heat Island Effects in Phoenix, Arizona. *Landsc. Urban Plan.* **2017**, *165*, 162–171. [CrossRef]
19. Meerow, S.; Newell, J.P. Spatial Planning for Multifunctional Green Infrastructure: Growing Resilience in Detroit. *Landsc. Urban Plan.* **2017**, *159*, 62–75. [CrossRef]
20. Liu, Y.; Wang, R.; Grekousis, G.; Liu, Y.; Yuan, Y.; Li, Z. Neighbourhood Greenness and Mental Wellbeing in Guangzhou, China_What Are the Pathways? *Landsc. Urban Plan.* **2019**, *190*, 9. [CrossRef]
21. Neema, M.N.; Ohgai, A. Multitype Green-Space Modeling for Urban Planning Using GA and GIS. *Environ. Plan. B Plan. Des.* **2013**, *40*, 447–473. [CrossRef]
22. Yuan, Z.; Tiemao, S.; Chang, G. Multi-Objective Optimal Location Planning of Urban Parks. In Proceedings of the 2011 International Conference on Electronics, Communications and Control (ICECC), Ningbo, China, 9–11 September 2011; pp. 918–921.
23. Maruani, T.; Amit-Cohen, I. Open Space Planning Models: A Review of Approaches and Methods. *Landsc. Urban Plan.* **2007**, *81*, 1–13. [CrossRef]
24. Thompson, C.W. Urban Open Space in the 21st Century. *Landsc. Urban Plan.* **2002**, *60*, 59–72. [CrossRef]
25. BenDor, T.; Westervelt, J.; Song, Y.; Sexton, J.O. Modeling Park Development through Regional Land Use Change Simulation. *Land Use Policy* **2013**, *30*, 1–12. [CrossRef]
26. Kim, H.; Tran, T. An Evaluation of Local Comprehensive Plans Toward Sustainable Green Infrastructure in US. *Sustainability* **2018**, *10*, 4143. [CrossRef]

27. Campagnaro, T.; Sitzia, T.; Cambria, V.E.; Semenzato, P. Indicators for the Planning and Management of Urban Green Spaces: A Focus on Public Areas in Padua, Italy. *Sustainability* **2019**, *11*, 7071. [CrossRef]
28. Lee, D.; Oh, K. The Green Infrastructure Assessment System (GIAS) and Its Applications for Urban Development and Management. *Sustainability* **2019**, *11*, 3798. [CrossRef]
29. Zhou, C.; Wu, Y. A Planning Support Tool for Layout Integral Optimization of Urban Blue–Green Infrastructure. *Sustainability* **2020**, *12*, 1613. [CrossRef]
30. Lai, S.; Leone, F.; Zoppi, C. Assessment of Municipal Masterplans Aimed at Identifying and Fostering Green Infrastructure: A Study Concerning Three Towns of the Metropolitan Area of Cagliari, Italy. *Sustainability* **2019**, *11*, 1470. [CrossRef]
31. Giles-Corti, B.; Broomhall, M.H.; Knuiman, M.; Collins, C.; Douglas, K.; Ng, K.; Lange, A.; Donovan, R.J. How Important Is Distance To, Attractiveness, and Size of Public Open Space? *Am. J. Prev. Med.* **2005**, *28*, 169–176. [CrossRef]
32. Liang, X.; Tian, H.; Li, X.; Huang, J.-L.; Clarke, K.C.; Yao, Y.; Guan, Q.; Hu, G. Modeling the Dynamics and Walking Accessibility of Urban Open Spaces under Various Policy Scenarios. *Landsc. Urban Plan.* **2021**, *207*, 103993. [CrossRef]
33. Chiesura, A. The Role of Urban Parks for the Sustainable City. *Landsc. Urban Plan.* **2004**, *68*, 129–138. [CrossRef]
34. Roe, M.; Mell, I. Negotiating Value and Priorities: Evaluating the Demands of Green Infrastructure Development. *J. Environ. Plan. Manag.* **2013**, *56*, 650–673. [CrossRef]
35. Lee, D.; Derrible, S. Predicting Residential Water Demand with Machine-Based Statistical Learning. *J. Water Resour. Plan. Manag.* **2020**, *146*, 04019067. [CrossRef]
36. Pedregosa, F.; Varoquaux, G.; Gramfort, A.; Michel, V.; Thirion, B.; Grisel, O.; Blondel, M.; Prettenhofer, P.; Weiss, R.; Dubourg, V.; et al. Scikit-Learn: Machine Learning in Python. *JMLR* **2011**, *12*, 2825–2830.
37. James, G.; Witten, D.; Hastie, T.; Tibshirani, R. *An Introduction to Statistical Learning with Applications in R*, 2nd ed.; Springer: Berlin/Heidelberg, Germany, 2013; Volume 112, ISBN 978-1-4614-7138-7.
38. Abhijith, K.V. Air Pollution Abatement Performances of Green Infrastructure in Open Road and Built-up Street Canyon Environments—A Review. *Atmos. Environ.* **2017**, *162*, 71–86. [CrossRef]
39. De Groot, R.S. *Functions of Nature: Evaluation of Nature in Environmental Planning Management and Decision-Making*; Wolters-Noordhoff BV: Toronto, ON, Canada, 1992; ISBN 90-01-35594-3.
40. Seoul Metropolitan Governments. *Urban Planning-Seoul's Unique, Sustainable Urban Planning*; Seoul Solution: Seoul, Korea, 2021.
41. OECD. *How's Life in Your Region? Measuring Regional and Local Well-Being for Policy Making*; OECD: Paris, France, 2014.
42. Byun, M.; Choi, J.; Park, M.; Lee, H. *The Quality of Life in Megacity and Seoul-Specific Happiness Indicator*; Seoul Research Institute: Seoul, Korea, 2015.
43. Bishop, C.M. *Pattern Recognition and Machine Learning*; Springer: New York, NY, USA, 2006; ISBN 978-0-387-31073-2.
44. Hastie, T.; Tibshirani, R.; Friedman, J. *The Elements of Statistical Learning*; Springer Series in Statistics; Springer: New York, NY, USA, 2009; ISBN 978-0-387-84858-7.
45. Lee, D.; Mulrow, J.; Haboucha, C.J.; Derrible, S.; Shiftan, Y. Attitudes on Autonomous Vehicle Adoption Using Interpretable Gradient Boosting Machine. *Transp. Res. Rec.* **2019**, *2673*, 865–878. [CrossRef]
46. Breiman, L. Random Forests. *Mach. Learn.* **2001**, *45*, 5–32. [CrossRef]
47. Friedman, J.H. Greedy Function Approximation: A Gradient Boosting Machine. *Ann. Statist.* **2001**, *29*. [CrossRef]
48. Lundberg, S.M.; Lee, S.-I. A Unified Approach to Interpreting Model Predictions. In Proceedings of the Advances in Neural Information Processing Systems; Guyon, I., Luxburg, U.V., Bengio, S., Wallach, H., Fergus, R., Vishwanathan, S., Garnett, R., Eds.; Curran Associates, Inc.: Red Hook, NY, USA, 2017; Volume 30, pp. 4765–4774.

 *sustainability*

Article

The Potential of Stormwater Management in Addressing the Urban Heat Island Effect: An Economic Valuation

Daniel Johnson *, Judith Exl and Sylvie Geisendorf

Environment and Economics, ESCP Business School Berlin, Heubnerweg 8-10, 14059 Berlin, Germany; judith.exl@edu.escp.eu (J.E.); sgeisendorf@escp.eu (S.G.)
* Correspondence: djohnson@escp.eu

Abstract: Urban green infrastructure (UGI) within sustainable stormwater management provides numerous benefits to urban residents, including urban heat island (UHI) mitigation. Cost–benefit analyses (CBA) for UGI have been conducted at neighborhood level with a focus on stormwater management, but valuations of reductions in heat-related hospitalizations and mortality are lacking. These benefits create significant social value; the quantification thereof is essential for urban planning in providing a scientific foundation for the inclusion of UGI in UHI mitigation strategies. This study assesses the potential of three UGI scenarios developed for an urban neighborhood in Berlin, Germany. First, climate data analyses were conducted to determine the cooling effects of tree drains, facade greening, and green roofs. Second, a CBA was performed for each scenario to value UHI mitigation by estimating the damage costs avoided in reduced heat-related hospitalizations and fatalities, using the net present value (NPV) and benefit–cost ratio (BCR) as indicators of economic feasibility. The results indicate heat mitigation capabilities of all three UGI types, with tree drains achieving the strongest cooling effects. Regarding economic feasibility, all scenarios achieve positive NPVs and BCRs above one. The findings confirm the potential of stormwater management in mitigating UHI and generating substantial social value.

Keywords: stormwater management; urban heat island; cost–benefit analysis; ecosystem services; urban green infrastructure

Citation: Johnson, D.; Exl, J.; Geisendorf, S. The Potential of Stormwater Management in Addressing the Urban Heat Island Effect: An Economic Valuation. *Sustainability* **2021**, *13*, 8685. https://doi.org/10.3390/su13168685

Academic Editor: Hyun-Woo Kim

Received: 15 July 2021
Accepted: 30 July 2021
Published: 4 August 2021

Publisher's Note: MDPI stays neutral with regard to jurisdictional claims in published maps and institutional affiliations.

1. Introduction

Urbanization combined with rising frequency and intensity of heat events as a result of climate change leads to increased occurrence of the so-called urban heat island (UHI) effect. This well-recognized phenomenon has been widely described in the literature and refers to increased ambient temperatures in heavily built-up urban environments as compared to rural areas [1]. The development of UHIs within cities are attributed to higher shares of sealed surfaces [2], for example, buildings and streets [3], and further exacerbate other urban issues such as stormwater management [4]. Moreover, building materials often absorb more heat during the day due to low solar reflectance, which is then emitted as thermal radiation during the night, reducing cooling phases between periods of daytime heat stress [5–7]. Average daytime UHI intensity recorded for European cities is currently 2.6 °C above temperatures in surrounding areas [8].

Besides the development of UHI as a result of urbanization, stormwater management becomes a pressing issue. Conventional solutions to stormwater management in dense urban areas attempt to convey stormwater quickly from the source to nearby water bodies or wastewater treatment plants [9]. Often, separated sewer systems are installed to convey the stormwater directly into nearby water bodies, carrying along polluting elements such as dust, heavy metals, and organic trace substances from construction and transportation [10]. Sustainable stormwater management is one alternative strategy to retain stormwater at the source and replicate natural processes in the water cycle, often referred to as sustainable urban drainage systems (SUDS) [11]. Such systems not only reduce stormwater runoff and

91

improve receiving river water quality [12] but also allow for the creation of numerous additional ecosystem services, which when monetarily valued provide substantial benefits that can outweigh the costs of installing and maintaining them [13]. With such a management approach, planners can instill multifunctionality in the planning process [14,15], addressing both stormwater management issues as well as numerous other issues such as the UHI effect [16]. In terms of mitigating UHI effects, measures within the framework of SUDS including tree drains (tree drains, or tree trenches, allow for the temporary storage and quicker infiltration of stormwater [17]), green roofs and façade greening provide substantial contributions and are further termed urban green infrastructure (UGI) [18].

UGI mitigates the UHI effect by offering heat regulation services through shading, evapotranspiration and thermal insulation of buildings [5,19–21]. Economic analyses of UGI have been conducted for individual measures, frequently with a focus on water management, e.g., [20–24]. Perini and Rosasco [25] also performed a monetary valuation of facade greening benefits such as heat regulation and air quality improvements. A complete analysis of UHI-related economic impacts on city scale has been carried out for Melbourne, Australia by Raalte et al. [26]. However, to date, the evidence of human health impacts of UGI in stormwater management is lacking [27], and only one study by Johnson and Geisendorf [13] assesses the economic performance of UGI measures on neighborhood level in Germany. The authors focus primarily on the valuation of ecosystem services ranging from water management to air quality improvement, but do not address heat regulation benefits provided by UGI. The purpose of the current study is to build on Johnson and Geisendorf's [13] study by quantifying the UHI mitigation effect of UGI for stormwater management and estimating the economic value of these temperature regulation benefits by identifying the damage costs avoided through reductions in heat-related hospital admissions and excess mortality. Three different UGI scenarios for a neighborhood in Berlin, Germany are compared to assess the potential of stormwater UGI to create net benefits for society by reducing urban dwellers' heat stress exposure. As the analysis is based on an actual management exercise within a research project that was developed for the Berlin study site, the calculation of UHI mitigation benefits can support the real planning process by providing quantitative information for objective evaluation of the different measures and scenarios and induce stakeholders to support the implementation of those UGI measures that create the highest net benefits for society.

2. UHI and Valuation

The UHI effect causes many negative impacts on urban life, resulting in high social and economic costs. Among the most commonly discussed effects are work productivity losses [28,29], increased energy consumption for indoor cooling [7,25], as well as increases in morbidity and mortality rates [5,30–32]. Heat-related mortality in particular represents a significant factor in UHI-related costs to society. Scherer et al.'s [32] comparison of approximately 1600 annual heat-related deaths in Berlin to an average of 64 road traffic fatalities in the German capital accentuates the topic's relevance and consequently the necessity for sustainable UHI mitigation strategies.

The effectiveness of UGI in lowering urban temperatures is well documented. Temperature regulation can be supplied by UGI mainly through evapotranspiration and shading, but also thermal insulation [8,20,21,33], and can be achieved in critical areas through tree drains supplemented with facade greening and green roof installation where space is limited. These measures not only positively impact the ambient air temperature [34] but also the indoor cooling loads [35], and the need for effects in urban areas have been apparent. For instance, Larondelle and Lauf [20] conducted a study investigating the supply and demand for different urban ecosystem services provided by UGI in Berlin, Germany. The demand for such services was found to be higher the more densely concentrated the infrastructure (e.g., in city centers). Moreover, Simperler et al. [36] found that areas in Berlin with high UHI correlated highly with areas also generating significant stormwater pollution and that UGI measures can combat both challenges at once. Such measures require substantial financial investments, underlining the need for objective economic valuation of non-marketable goods such as heat regulation, in order to create transparency in policy making [37]. Although arguments against economic valuation as a comprehensive approach to encompassing the many facets of nature's value have their legitimacy [38,39], we see the importance of including monetary valuation of ecosystem services in enabling cost-efficient urban planning processes [40].

As the present analysis incorporates private and social elements, ecosystem services and health-related impacts over time, a brief discussion on discounting is warranted. We adopt a dual discounting scheme to discount private and social costs and benefits separately [41]. Although there are ethical concerns regarding the use of exponential discount rates [42], we do not pursue a hyperbolic discounting scheme since the impacts are analyzed over a shorter time horizon than typical climate-economic models that extend beyond the end of the current century [43]. The 3% discount rate for private costs and benefits is the same rate used in the KURAS project [44] and suggested by the German Association for Water, Wastewater and Waste for use in German economic analyses [45]. We use a lower 2.1% discount rate for ecosystem services accruing as social benefits, in light of recent research on the proper discount rates to be used for ecosystem services [46,47]. This rate is also used for health-related impacts, which resembles the rate suggested by Attema et al. [48]. However, it should be noted that there are grounds for discounting health-related impacts with a zero-discount rate, as society may place little to no difference in weight on future lives saved [49,50].

3. Materials and Methods

This study assesses the economic value of reductions in heat stress by three UGI scenarios on a neighborhood level while incorporating the monetary valuation of ecosystem services. First, the analysis is based on climate data from the KURAS project ("Concepts for urban stormwater management and sewer systems"; German: "Konzepte für urbane Regenwasserbewirtschaftung und Abwassersysteme", http://www.kuras-projekt.de/ (accessed on 17 July 2017)) [12], funded by the German Federal Ministry for Education and Research. The project modeled an urban neighborhood within the district of Pankow, Berlin, which is the site for current investigation. The second part of the analysis concerns the economic valuation of the results obtained from the climate data analysis. The UHI mitigation effects of UGI for the three scenarios are combined with demographic statistics as well as heat-related morbidity and mortality data from the literature to calculate UHI mitigation-related benefits. Other UGI benefit and cost calculations for the Berlin study site were also taken from existing literature to supplement the UHI mitigation benefit data.

3.1. Study Site

The Berlin neighborhood is a residential area of 117 hectares within the urban district of Pankow, characterized by a combination of multi-story apartment buildings from the 1920s and 1930s, and higher residential buildings from the post-war period [44]. In terms of demographic indicators such as population density and age structure, it represents a typical Berlin neighborhood. Although the KURAS project focused on stormwater management and therefore included measures such as swales and ponds, the UGI measures relevant for this study were part of the project scope, and their effects could be isolated for the purpose of the benefit calculations. Moreover, we align the economic valuation of ecosystem services for the three UGI scenarios designed in the KURAS project according to the methodology described in Johnson and Geisendorf [13].

The KURAS project partners developed three independent scenarios for the implementation of urban stormwater management measures, including UGI, based on local requirements—for example, improving groundwater management, urban climate, and biodiversity—as well as stakeholder priorities and feasibility [44]. The creation of different scenarios allows identification of the most efficient UGI placement, which is an important factor because the strategic implementation of UGI measures in key areas yields better results than simply increasing the share of green cover [21]. Furthermore, the scenarios were developed independently of one another within separate groups of stakeholders, planners and interest groups. Figure 1 shows the placement of UGI measures as part of the stormwater management scenarios.

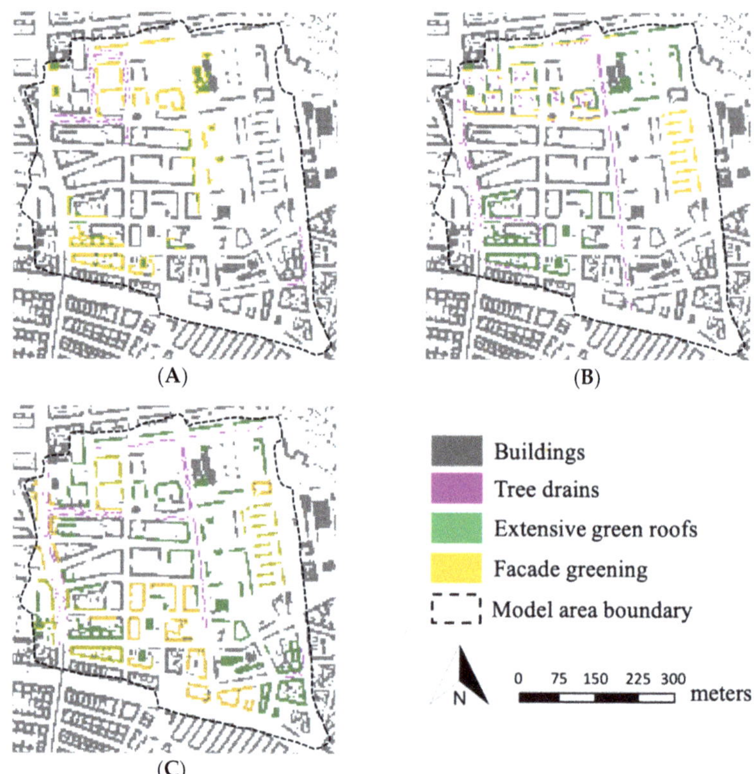

Figure 1. Placement of UGI measures for stormwater management scenarios (**A**–**C**).

Whereas Scenario A is characterized by the lowest coverage of greening, Scenario C is characterized by the highest overall greening coverage. In particular, Scenario C receives more façade greening and extensive green roofs than the other two scenarios, and the level of tree drain coverage is only slightly lower than Scenario B. Although Scenario B has a higher coverage of tree drains and extensive green roofs than Scenario A, the application of façade greening is greatly lower than in the other two scenarios, which significantly lowers the overall costs as seen below in Section 4.2.

3.2. Climate Data Analysis

The KURAS climate data that were evaluated for the current study is organized in a Cartesian coordinate system with 8 m × 8 m grid cells. A value attached to each grid cell represents how the specific space is being used, that is, if the area is covered with buildings, streets, other (partially) sealed surfaces such as parking lots or railways, or if green or blue spaces such as grass, trees or ponds are present. Additionally, climate data for each grid cell is available with temperatures measured at a height of 2 m (i.e., pedestrian level), showing diurnal heat stress measured in hours per year. For categorization of daytime heat stress, the universal thermal climate index (UTCI) was selected, including UTCI classes strong heat stress (SHS; >32 °C–36 °C), very strong heat stress (VSHS; >36 °C–46 °C), and extreme heat stress (EHS; >46 °C). This selection is consistent with the German Meteorological Service's (German: "Deutscher Wetterdienst", DWD) policy of issuing heat warnings above temperatures of 32 °C [51]. Daytime heat stress, that is, UTCI in hours per year, for the current situation in the Pankow study site is shown in Figure 2.

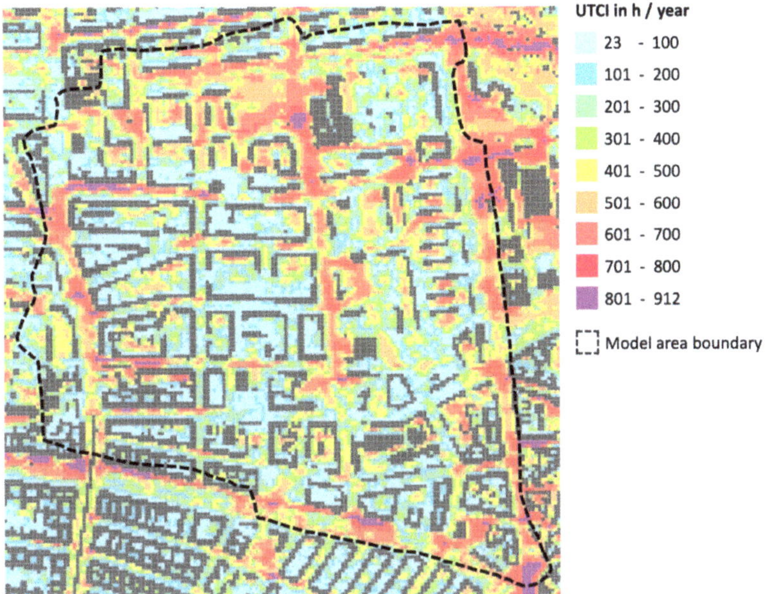

Figure 2. Heat stress in the current situation for the study site in Pankow, Berlin.

For all three scenarios, the grid cell values show where stormwater management measures were added. These measures include extensive green roofs, facade greening, urban trees, swales, swale-trench systems, swale-trench beds, ponds, permeable pavement, rainwater harvesting cisterns, and retention soil filters. R software [52] was used to filter out only those grid points where UGI (i.e., tree drains, facade greening, or green roofs) was added. Temperature differences were then calculated for these grid points as compared to the base scenario, specifically extracting minimum, maximum and mean temperature changes achieved by each UGI type, expressed in UTCI.

3.3. Economic Valuation

UHI mitigation effects achieved by UGI were valued in monetary terms. The damage costs avoided by lowering heat-related morbidity and mortality were combined with additional ecosystem services following the methodology of Johnson and Geisendorf [13]. Once quantified, those benefits were balanced against initial UGI installation costs as well as recurring maintenance and operation expenses taken from Strehl et al. [53]. In the final step, the resulting net benefits were calculated over a time span of 50 years and subsequently discounted to present value in order to obtain the NPV and BCR.

The first step in the monetization of UHI mitigation benefits was to calculate the share of the study site area covered with additional UGI in each of the three scenarios developed by the KURAS project. Next, the total heat stress reduction per scenario was calculated as the sum of all UTCI classes (i.e., SHS, VSHS, and EHS) and converted from hours per year to the number of hot days per year. This conversion was performed assuming an average of 16.03 h of daylight in Berlin for the months of June–August [54], which is the period during which most heat events and the majority of heat-related fatalities occur [32,55].

Demographic data from the German Federal Statistical Office (German: "Statistisches Bundesamt") and Umweltatlas Berlin [56] were used to determine the number of residents that are reached by UGI measures in the Pankow neighborhood and consequently benefit from reduced heat stress and lower morbidity and mortality rates. The calculation method was modeled after a similar approach used by Bodnaruk et al. [19]: population data was first calculated on neighborhood level and then multiplied by the percentage of the model site area covered with UGI to obtain the number of residents who actually benefit from the additional green cover. Subsequently, the numbers of these citizens falling into the 0–74 years and 75+ years age groups were calculated based on the respective shares of 92.16% and 7.84% [57], since the latter age group has been found to be significantly more vulnerable to heat-related morbidity and mortality increases [28]. Different excess morbidity and mortality rates were consequently applied to both age groups. All demographic data used in the UHI mitigation benefit calculations is for the year of 2017. Using values from this single year rather than an average from a range of years was deemed appropriate because 2017 can be classified as a statistically average weather year: DWD climate summary reports for the years 2014–2018 reveal that all five years showed average annual temperatures of 1.4–2.2 °C above the international reference period of 1961–1990. The value for 2017 was 1.4 °C above average and no unusual number of heat events occurred during 2017 compared to the other years [58–62].

3.3.1. Reduced Heat-Related Morbidity

The demographic situation for the study site base scenario is as follows: According to Umweltatlas Berlin [56], the total population of the Pankow study site was 23,276 in 2017. A total of 21,451 of this number fall into the 0–74 age group with a 17.09% annual hospitalization rate for all causes, and the remaining population of 1825 are aged 75 years or older with a base hospital admission rate of 49.86%, consistent with hospitalization statistics for Germany [63].

Reductions in heat-related hospitalization cases by age group were calculated for each scenario using increases in morbidity rates for UTCI class SHS based on a study conducted by Johnson et al. [64] for the 2003 heat wave in England, and EHS morbidity rate increases as calculated by Hübler et al. [28] based on the data provided by Johnson et al. [64]. Excess morbidity for VSHS was not considered by either study, therefore linear extrapolation of moderate heat stress and SHS values from Johnson et al. [64] was used to determine the effect for VSHS. Total excess morbidity rates obtained in this manner amount to 1.82% for SHS, 2.91% for VSHS, and 3.99% for EHS. These values were then converted to a weighted average increase for hot days (i.e., days with average temperatures above 32 °C) by UTCI class incidence in the current situation, yielding relative morbidity increase rates of 0.76% for the 0–74 age group, and 16.93% for the 75+ age group. These rates were subsequently multiplied by the share of the model site population reached by UGI measures and applied to the reduction in hot days achieved by each UGI scenario as calculated from KURAS climate data.

Exact quantification of the costs related to heat-related hospitalizations is challenging as no data is available on treatment costs of heat-related illnesses in Germany [28,65]. Therefore, Hübler et al. [28] carried out an economic appraisal of the overall expenditures based on average hospitalization costs in Germany. According to the authors' calculations, current heat-related hospitalization costs amount to EUR 82 million and will increase six-fold for the period 2071–2100, taking into account climate change as well as demographic changes such as population size and age structure. In relative terms, this figure represents 0.27% of total German healthcare expenses. Each prevented hospital admission case was estimated to account for an average of EUR 3300 in damage costs avoided, with the estimate referring to statistics on hospital treatment costs for heat-related illnesses [28,65].

We derive the estimated economic value of the reduced heat related morbidity (V_{morb}) in accordance with the damage cost avoided for age groups k (0–74 years) and K (75+ years):

$$V_{morb} = \sum_{k}^{K} Pop_{UGI,k} \cdot Morb_{2017,k} \cdot Morb_{heat,k} \cdot \frac{UTCI_{red,trees} + UTCI_{red,GR} + UTCI_{red,FG}}{365 \; days} \cdot HC_{avg} \tag{1}$$

where $Pop_{UGI,k}$ is the number of residents reached by UGI in the specific age group (0–74 years: 4130 residents; 75+ years: 351 residents) [56], $Morb_{2017,k}$ is the base morbidity rate in 2017 for the specific age group (0–74 years: 17.09%; 75+ years: 49.86%) [63], $Morb_{heat,k}$ is the relative increase in heat-related morbidity rate for the specific age group (0–74 years: 0.76%; 75+ years: 16.93%) [28,64], $UTCI_{red}$ is the average reduction in UTCI exceedance in hours per year of cells by UGI (trees drains, *trees*; green roofs, *GR*; façade greening, *FG*), and HC_{avg} is the average hospitalization cost.

3.3.2. Reduced Heat-Related Mortality

Mortality cases in the current situation were calculated using the base mortality rate of 1.04% for Germany [66]. The base mortality rate was adjusted for seasonality, taking into account that winter mortality has been found to be 8% higher and summer mortality 8% lower than the annual average [28]. Reduced mortality for each scenario was determined by applying a heat-related excess mortality rate of 5% [32] to the base mortality rate and multiplying the result with the number of model site residents reached by UGI cooling effects, as well as the reduction in number of hot days as calculated from UTCI data (Table 3). Based on data from the 2003 European heat wave, 20% of the mortality reduction was attributed to the 0–74 age group, while the remaining 80% were assumed to occur in residents aged 75 years and older [28]. In order to express the value of these avoided deaths in monetary terms, population percentages by age were used to break down the average life expectancy in Germany of 78.81 years to the average number of life years lost per death, yielding 49.88 years for the 0–74 age group and 5.76 years for the 75+ age group [57,67,68].

Each lost year of life was valued at EUR 95,653 based on dividing the VSL for Germany (EUR 7.538 million), as estimated by Viscusi and Masterman [69], by the current average life expectancy. Viscusi and Masterman [69] provide the most up to date VSL by taking a base VSL derived through a meta-analysis of revealed preference data from hedonic wage studies and income elasticity for the U.S. and estimating country-specific VSLs. Although the OECD [70] proposes a base VSL of approximately USD 3.6 million (2005 dollars), this estimate is based on stated preference data, whereas Viscusi and Masterman [69] discuss the disadvantages of the OECD [70] values and base their data on revealed preferences to mitigate hypothetical bias. In order to account for uncertainty in such a large VSL, we test for sensitivity of the cost–benefit model to the VSL by incorporating the Germany-specific VSL estimate from Spengler [71] of EUR 1.978 million as a minimum value.

We derive the estimated economic value of the reduced heat related mortality (V_{mort}) in accordance with the damage cost avoided for age groups k (0–74 years) and K (75+ years):

$$V_{mort} = \sum_{k}^{K} Pop \cdot M_{2017} \cdot M_{heat,k} \cdot d_{season} \cdot UGI \cdot \frac{UTCI_{red}}{UTCI_{current}} \cdot \left(LE_k - AGE_{avg,k} \right) \cdot VOLY \tag{2}$$

where Pop (23,276 residents) is the population in the study site in 2017 [56], M_{2017} (1.13%) is the base mortality rate of Germany in 2017 [66], $M_{heat,k}$ is the excess heat-related mortality rate (5%) in Berlin [32] given the 0–74 age group (4%) and 75+ age group (1%) [28], d_{season} is the adjustment for seasonal mortality (i.e., summer mortality 8% lower and winter mortality 8% higher than annual average [28]), UGI is the share of the study site area (%) to be complemented with UGI, $UTCI_{red}$ is the average reduction in UTCI exceedance in hours per year of cells with UGI, $UTCI_{current}$ is the current average UTCI exceedance in the cells with UGI, LE is the average life expectancy for the specific age group, AGE_{avg} is average age of residents in the age group, and $VOLY$ is the value of life year derived from the average life expectancy (78.8 years [66]) and the inflation-corrected VSL of EUR 7.538 million [69].

3.3.3. Ecosystem Service Valuation

The UGI measures in this study bring about numerous additional ecosystem services. The methodology for economic analyses of these ecosystem services is described by Johnson and Geisendorf [13], and the applicable results from their research were included in the CBA performed for this study.

The ecosystem services valued for UGI in this study included reduced stormwater runoff [10], increased building longevity [72], habitat creation [73], increases in building value (aesthetic improvements) [74], heating and cooling savings [72], and pollution [75] and carbon dioxide (CO_2) reduction [25]. For example, green roofs and tree drains directly contribute to reducing loads on municipal treatment plants. Berlin charges a rainwater fee for sealed surface at 1.804 EUR/m^2 for home and property owners, but this fee can be reduced by 50% for such UGI measures [76]. We assume that tree drains provide additional stormwater retention for 250 m^2 area on average.

The remaining relevant benefits could be completely attributed to UGI implementation and were therefore computed directly according to Johnson and Geisendorf [13]: longevity increases apply to green roofs and facades, which can be replaced or restored at larger intervals compared to conventional roofs and facades due to the building materials' protection against weathering by greening. Habitat creation benefits are added by green roofs and could therefore also be transferred entirely. The same was true for aesthetic improvements achieved by facade greening, heat savings and avoided heat externalities resulting from green roof installation, as well as cooling effects of green roofs and facades. Air quality improvement calculations were equally based on UGI effects only. They include pollution removal by all UGI types, and carbon storage and sequestration by tree drains and green roofs.

3.4. Cost–Benefit Model

Unit investment costs as well as operation and maintenance costs for the three UGI types were taken from an interim report presenting the economic side of the KURAS project [53], as well as Johnson and Geisendorf [13], and used to calculate total costs of UGI measures implemented in each scenario. The unit costs for each measure are shown in Table 1. Tree drains are clearly the most expensive UGI type to install, but operation and maintenance costs are lower than for the other two measures. Green roofs are a low-cost choice both in terms of installation and running costs, while facade greening generates the second-highest installation costs as well as the highest annual costs.

Table 1. Unit installation (I)/annual operation and maintenance (O&M) costs of the three UGI types.

	I (EUR/m^2)	O&M (EUR/m^2)
Extensive green roofs	20	1.50
Façade greening	100	10.00
Tree drains	303	1.21

Benefits gained from lowering morbidity and mortality rates as a result of reduced heat stress were added to the additional ecosystem services. The economic analysis was calculated over a project horizon of 50 years, reflecting the typical lifetimes of UGI measures: green roofs are estimated to need replacement after 40–55 years, green facades after 40–45 years, and trees can remain in place for several decades beyond this time horizon [25,72]. Both the NPV and BCR were selected as indicators for economic comparison of the three scenarios. When the NPV of a project is greater than zero, or the BCR greater than 1, this indicates that the present value of all benefits outweighs the sum of the discounted costs over the project period, and pursuing the project is an economically sound option. Future costs and benefits were discounted to present value following the discounting regime described in Section 2.

3.5. Sensitivity Analysis

A sensitivity analysis was performed to test the sensitivity of the results to changes in the parameters of cost–benefit model. According to the European Commission [77], if a 20% change in a project parameter leads to a 20% change in the NPV, the NPV is assumed to be sensitive to the parameter. Johnson and Geisendorf [13] demonstrated that the NPV may be sensitive to the installation and maintenance costs as well as the property value increases due to green roofs. Given the VSL and discount rate may produce significant effects on the NPV, we additionally included these parameters in the sensitivity analysis.

The sensitivity analysis was performed in the form of a Monte Carlo analysis with 10,000 iterations of repeated drawings from distributions of the parameters for which the NPV can be sensitive. We varied the parameters according to distributions in Table 2.

Table 2. Input parameters and distributions for the Monte Carlo simulations with U(minimum, maximum) as a uniform distribution, N(mean, standard deviation) as a normal distribution and T(minimum, maximum) as a triangular distribution.

	Distribution and Parameters	**Units**
Installation costs	U(−20, +20)	%
Operation and maintenance costs	U(−20, +20)	%
Property value increase in green roofs	U(2, 5)	%
Heat-related mortality rate	N(5.2, 1.3)	%
Value of statistical life	T(1,978,165, 7,538,485)	EUR
Private discount rate	U(1, 5)	%
Social discount rate	U(0, 3)	%

Since the installation and operation and maintenance costs of the scenarios were partially determined through general and site-specific cost estimations, we assume a range of costs at ±20% for each of the scenarios. For the property value increases in green roofs, we assume an increase in building values in the range from 2–5% based on the study by Bianchini and Hewage [74]. For heat-related mortality, we follow the mean and standard error values given in Scherer et al. [32]. The VSL follows a triangular distribution because the Viscusi and Masterman [69] estimate is the most likely (also the maximum), although we acknowledge prior studies [71] estimating the Germany-specific VSL be much lower (minimum). The range of private discount rates used follow those tested in Johnson and Geisendorf [13], and the social discount rates are assumed to follow a uniform distribution but only maximizing to the private discount rate for a given iteration of the analysis.

4. Results

A visual representation of the UHI mitigation in each of the three scenarios can be seen in Figure 3. Results of the climate data analysis including minimum, maximum and mean reductions in UTCI by UGI measure and for each scenario are summarized in Table 3.

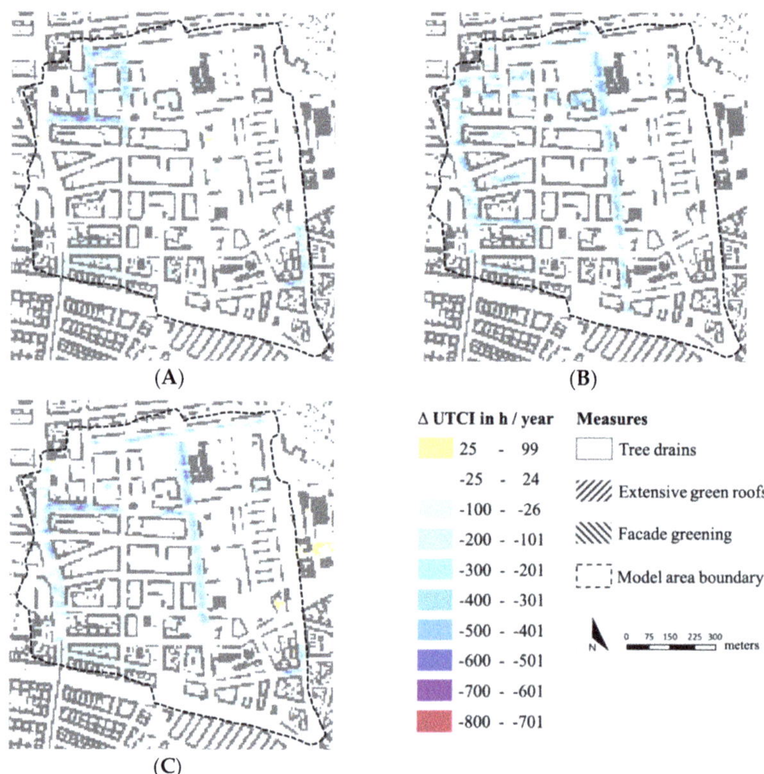

Figure 3. Changes in heat stress as reduced hours of UTCI exceedance for each of the scenarios, (**A–C**) is a visual representation of the UHI mitigation in each of the three scenarios.

Table 3. Changes in UTCI following the implementation of UGI measures for the three scenarios.

Scenario	UGI Measures	Areas	Change in UTCI (hours/year)		
			Min	Max	Mean
A	Tree drains	5341	−711.49	0.00	−319.52
	Extensive green roofs	58,515	−380.80	96.10	−10.94
	Façade greening	199,880	−380.80	96.10	−12.67
B	Tree drains	10,631	−593.14	0.00	−221.59
	Extensive green roofs	72,926	−316.5	19.18	−8.98
	Façade greening	79,008	−76.78	10.34	−3.01
C	Tree drains	10,118	−669.00	0.00	−285.71
	Extensive green roofs	121,658	−297.40	37.84	−4.68
	Façade greening	290,732	−230.21	22.08	−5.73

Comparing the effects of the different UGI measures, tree drains generally achieve the highest mean daytime heat stress reductions, whereas facade greening and green roofs perform at similar, lower levels across the three scenarios. Using mean heat stress reduction values and assuming an average of 16.03 daylight hours per day for the heat event season from June to August in Berlin, Scenario A experiences 21.41 fewer hot days compared to the current situation; Scenario B 14.57 days; and Scenario C 18.48 days.

4.1. Economic Valuation

The UGI benefits are shown in Table 4, with all benefits calculated over the project time horizon of 50 years and discounted at 3% for private benefits and 2.1% for social benefits.

Table 4. Present value (EUR) of all benefits over the 50-year time horizon discounted at 3% for private benefits and 2.1% for social benefits. Benefits provided by tree drains (TD), extensive green roofs (GR), and façade greening (FG). Valuation methods include damage cost avoided (DCA), market price (MP), replacement cost (RC), and benefit transfer with hedonic pricing (BT).

	Benefits	Provided by	Valuation Method	Type of Benefit	Scenario A	Scenario B	Scenario C
UHI Mitigation	Heat-related mortality reduction	TD GR FG	DCA	Social	35,057,726	20,436,515	34,460,982
	Heat-related morbidity reduction	TD GR FG	DCA	Social	208,386	95,900	270,181
Ecosystem services	Runoff reduction	TD GR	MP	Private	2,309,569	3,520,134	4,575,692
	Increasing building longevity						
	Roof longevity	GR	RC	Private	1,732,711	2,159,438	3,602,475
	Façade longevity	FG	RC	Private	8,985,805	3,551,884	13,070,148
	Habitat creation	GR	RC	Social	474,941	591,909	987,449
	Aesthetic improvements						
	Property value (w/façade greening)	FG	BT	Private	14,038,596	8,896,971	18,522,905
	Property value (w/green roof)	GR	BT	Private	20,050,658	24,988,683	41,687,285
	Energy savings						
	Heating savings	GR	MP	Private	1,842,809	2,296,651	3,831,380
	Cooling savings	GR FG	RC	Social	462,488	474,455	886,747
	Heating externalities	GR	DCA	Social	1,098,200	1,368,662	2,283,265
	Air quality improvements	TD GR FG	DCA	Social	392,520	311,150	682,411
	CO_2 storage and sequestration	TD GR	DCA	Social	112,187	157,210	228,933
	Totals				86,766,596	68,849,562	125,089,853

Regarding UHI mitigation benefits, heat-related hospitalizations are reduced by 2.05 cases per year for Scenario A, by 0.94 for Scenario B, and by the highest number of 2.66 cases for Scenario C. Heat-related morbidity reductions result in comparatively low monetary benefits, with Scenario C reaching the highest value of EUR 270,181 in damage costs avoided over the project horizon. In terms of mortality reduction, the results for Scenarios A and C are almost identical with 0.78 and 0.77 annual deaths avoided, respectively. For Scenario B, heat-related mortality cases are reduced by 0.46 per year, resulting in lower societal benefits for that scenario.

Combining both morbidity and mortality reductions, the results for both Scenarios A and C are similar, with the benefits of Scenario A being slightly higher at EUR 35,266,112 compared to EUR 34,731,163 for Scenario C. Scenario B achieved considerably lower UHI mitigation benefits of EUR 20,532,416. The similar performances of Scenarios A and C may initially seem surprising, particularly in view of the fact that Scenario C receives a much higher portion of additional green cover—28.92% of the entire study site area receive greening in this scenario, as compared to 19.25% for Scenario A. This means that a larger share of the population benefits from heat mitigation through UGI, which should result in higher UHI mitigation benefits for Scenario C. However, at the same time the mean heat stress reduction effect of UGI measures implemented in Scenario A is larger

(Table 3). The higher effectiveness seems to be due to particularly successful placement of UGI measures in areas affected by exceptionally strong heat stress. This finding is in line with the conclusion of Zölch et al. [21] who, as mentioned in the literature review, note that placing UGI in strategically advantageous areas is more important than aiming for sheer increase in volume.

When combining both UHI mitigation benefits and ecosystem services, the resulting total UGI benefits of the three scenarios were as follows: Scenario C performed best with EUR 125,089,853; even though UHI mitigation benefits are comparable with those of Scenario A, the other benefits tip the balance in total benefit calculations. Scenario A follows with considerably lower total benefits of EUR 86,766,596, and Scenario B amounted to EUR 56,831,176. UHI mitigation benefits contributed a share of 41%, 30%, and 28% to the total benefits of Scenarios A, B, and C, respectively, with only aesthetic improvements achieving higher contributions for Scenarios B and C. This identifies heat regulation as one of the main contributing factors in UGI benefit generation.

4.2. Cost–Benefit Analysis

Comparing the three scenarios' costs as calculated over the entire project time horizon (Table 5), it is clear that Scenario C is the costliest to implement and maintain with total costs for installation as well as operation and maintenance. These high costs can be explained by the much higher area to be covered with additional UGI measures for Scenario C as compared to the other two scenarios, particularly with regard to facade greening, which generates the highest operation and maintenance costs of all UGI measures (Table 1).

Table 5. Present value of all costs and installation (I) and annual operation and maintenance (O&M) costs of the three scenarios [54].

	Scenario A		Scenario B		Scenario C	
	I (EUR/m^2)	M (EUR/m^2)	I (EUR/m^2)	M (EUR/m^2)	I (EUR/m^2)	M (EUR/m^2)
Tree drains	1,618,348	5988	3,221,973	11,921	3,065,945	11,344
Extensive green roofs	1,170,294	87,772	1,458,511	109,388	2,433,156	182,487
Façade greening	19,988,560	1,998,856	7,900,800	790,080	29,073,220	2,907,322
Totals	22,777,202	2,092,616	12,581,284	911,389	34,572,321	3,101,153
Present value of all costs	76,619,718		36,031,108		114,364,256	

Table 6 shows the NPVs and BCRs of the three scenarios, combining both discounted costs (Table 5) and discounted benefits (Table 4) linked to each scenario. Scenario B clearly performed the best given the higher BCR and NPV. Despite greatly different implementation of UGI and corresponding costs, the performances of Scenarios A and C are very similar in both the NPV and BCR calculations and make the cutoff for economic feasibility. The significantly better performance of Scenario B despite its comparatively low benefit contribution is largely due to the fact that implementation costs for this scenario are by far the lowest. Scenario B included a high share of additional trees with tree drains, which are the costliest UGI type in terms of installation However, tree drains are also the most inexpensive measure when it comes to operation and maintenance costs (Table 1), lowering the overall cost over the project life span. Scenario B also receives the lowest share of facade greening, which is the costliest measure in terms of operation and maintenance. In sum, these low costs compensate for the relatively low benefits to be gained from implementing Scenario B.

Table 6. Net present values (NPV) and benefit–cost ratios (BCR) of the three scenarios.

	Scenario A	Scenario B	Scenario C
NPV	EUR 10,146,769	EUR 32,818,400	EUR 10,725,544
BCR	1.13	1.91	1.09

The similar performances of scenarios A and C, even with Scenario C receiving a significantly higher share of UGI, can be traced back to the more efficient UGI placement for Scenario A mentioned above. This results in almost the same level of UHI mitigation benefits for both scenarios, which combined with lower costs for Scenario A ultimately leads to the slightly better performance of this scenario. In other words, Scenario A achieves the same level of UHI mitigation benefits at a much lower cost, meaning that benefits are reached in a more cost-effective manner. Nevertheless, both Scenarios A and C have low BCRs, which becomes a relevant factor in the sensitivity analysis presented in 5.3. This leaves Scenario B as the most robust option; while the scenario generates lower benefits than the other two, the discounted costs are also substantially lower, resulting in a higher positive net outcome and thus confirming that this scenario generates substantial value for society.

4.3. Sensitivity Analysis

A Monte Carlo analysis was performed with 10,000 draws from the given distribution of parameters in Table 2 to test the sensitivity of the NPV to the parameters. The number of draws is assumed to be amply large to obtain a random sample from each distribution with all possible combinations. We obtained probabilities of achieving an economically feasible result (NPV > 0) of 64%, 100%, and 63% for Scenarios A, B, and C, respectively. Therefore, the results are robust given the uncertainty of the tested parameters. Given that Scenario B would remain positive even with a significantly lower VSL estimate, we achieve a much higher likelihood of economic feasibility.

We also tested the probabilities of achieving the baseline NPV, which resulted in probabilities of 34%, 39%, and 40% for Scenarios A, B, and C, respectively. The triangular distribution of the VSL became a significant factor for these probabilities because the maximum VSL was already the estimate used for the baseline calculation. Additionally, the uniform distribution for a ±20% change in total installation and operation and maintenance costs lead to higher risks in terms of the likelihood of achieving a positive outcome. However, we maintain that the cost estimates provided by Strehl et al. [53] are already general costs estimates from the literature and expert estimations, and a 20% variation from these would likely gravitate to more extreme outliers.

5. Discussion

Statistical climate data analysis and CBA were used as methods for determining the effectiveness of three stormwater management scenarios developed for a neighborhood in Pankow, Berlin, in combating UHI effects and thereby creating value for society. The findings indicate two things: first, the climate data analysis reveals mean temperature reductions in locations where UGI measures were added, confirming that UGI measures in stormwater management can be effective in mitigating UHI effects. Secondly, lower morbidity and mortality rates resulting from this decreased heat stress created monetized benefits which contributed significantly to the positive economic performance of the three scenarios, allowing the conclusion that UGI measures can be implemented in an economically feasible manner.

The climate data analysis showed that UGI can be effectively employed to mitigate urban heat stress, with varying results depending on UGI type as well as strategic placement of measures. The results for all three scenarios generally match the findings of previous studies. For tree drains, the climate data analysis showed the highest daytime heat stress reductions of up to 711 h per year (Table 3). This is in line with the findings of existing literature, where trees are found to have the strongest heat mitigation effect which can even extend to several meters distance from the trees themselves [8,21,78]. Facade greening, by contrast, achieved considerably lower cooling effects; a maximum of 381 h in annual heat stress reductions could be confirmed. This pattern is also found in existing studies which are in general agreement that green facades contribute less to heat stress mitigation for residents, as the effect seems to be limited to the immediate proximity of the walls [21,33,79]. One finding that could not be confirmed by the climate data analysis is the indoor cooling effect of facade greening, as the climate data were measured and simulated for outdoor areas only in the KURAS project. Green roofs reached cooling effects very similar to those of green facades, with the same maximum daytime heat stress reduction of 381 h per year. Existing literature offers mixed findings for the climate effects of green roofs, with some studies not finding any effect at street level, and others acknowledging very slight effects [7,8,21,31,65]. The effects that are revealed by the climate data analysis are mostly due to low roofs such as underground parking lots, but overall the UHI mitigation contribution of green roofs is low during the day [44].

Given the high heat-related benefits of the study, it is important to consider plausibility as few studies have empirically investigated the green cover and improved health outcomes relationship, i.e., [80]. The UHI mitigation benefits that were calculated for the study site, that is, reductions in heat-related hospitalizations and deaths, were tested for plausibility by extrapolating the CBA results to national level and comparing them with actual heat-related hospitalization and mortality statistics for Germany. For heat-related hospitalizations in Germany, 1856 cases were recorded in 2013, while 2003, the year of the European heat wave, resulted in 2561 hospitalizations [81]. Taking the maximum number of hospitalization cases avoided for the Pankow, Berlin study site, hospitalizations were reduced by 2.66 cases for a total of 23,276 residents. Linearly extrapolating this to a population of roughly 80 million for all of Germany yields a reduction of 9142 cases per year. This figure evidently exceeds the number of regular countrywide hospitalization cases by a factor of 4. Possible reasons for this could be that due to the UHI effect, urban dwellers are at a higher risk of suffering heat-related illnesses, leading to a higher share of hospitalizations in the city compared to the national average. Conversely, building resilience against heat stress through UGI has a greater impact and reaches a greater number of citizens in a more efficient manner in densely populated urban areas, which could also lead to the comparatively high number of avoided hospitalization cases in the model site area. Nonetheless, assuming that the benefits could indeed have been overestimated, this would be counterbalanced to a degree by a conservative hospitalization cost estimate. Calculating total treatment costs for Germany for 9142 heat-related hospitalizations at EUR 3300 per case—the amount used in the CBA— we arrive at just over EUR 30 million in annual costs, representing less than 40% of the EUR 82 million in national heat-related hospitalization costs [28]. Moreover, the economic effect of reduced hospital admissions on the CBA calculations remains in any event limited, as these savings account for only 1% of total UHI mitigation benefits.

Annual heat-related mortality cases for Germany have been estimated at 4500 deaths [65]. Taking the maximum number of deaths avoided for the model site area, we arrive at 0.78 cases for a population of 23,276, which would correspond to 2681 deaths avoided at national level. This equals only approximately 60% of the number of heat-related deaths in Germany estimated by Tröltzsch et al. [65]. Assuming that, as with hospitalizations, the occurrence of heat-related fatalities is not evenly distributed throughout the country, but that incidence is clustered in urban areas, the results appear to be even more conservative estimates. However, we tread lightly in this conclusion as there is evidence of urban-rural disparities in heat-related mortality, with higher rates being found in rural areas [82]. Despite this evidence, the outcome is consistent with observations made in the literature review of Berlin accounting for a disproportionate share of Germany's heat-related deaths. Looking at the economic valuation of heat-related deaths, Tröltzsch et al. [65] arrive at an inflation-adjusted annual cost to society of EUR 2.24 billion for 4500 deaths, assuming an average of 8 lost life years per death. Using the number of fatalities as calculated for the Pankow study site and extrapolating it to national level, total heat-related deaths in Germany amount to 2681 cases. Applying the same VOLY of 62,126 as calculated by Tröltzsch et al. [65] as well as a slightly higher average of 9.56 life years lost per case, total cost savings would amount to EUR 1.59 billion, or 70% of the total costs calculated by Tröltzsch et al. [65], suggesting conservative benefit estimates for this study.

Overall, the results show that by addressing stormwater management issues, specific UGI measures can be used as a cost-effective method for lowering urban heat stress, but at the same time the strategic selection and placement of measures is key to achieving economic feasibility. This becomes most evident when comparing the shares of green cover and temperature reduction effects for Scenarios A and C, with Scenario C receiving by far the highest share of UGI, but Scenario A still performs better in terms of heat mitigation due to more effective placement of the measures.

6. Conclusions

Urban areas are increasingly affected by stormwater management issues and the UHI effect due to a continuing urbanization trend leading to growing urban populations and further consolidation of already heavily built-up city centers. The implementation of UGI measures encompassed in sustainable stormwater management represents one solution to mitigating heat stress for the urban population while also providing numerous additional benefits. The purpose of this study was to quantify and value the UHI mitigation potential through heat-related mortality and morbidity reductions in three different UGI scenarios and assess the economic feasibility using a CBA approach. Two contributions were made to the literature on valuing UGI benefits in a UHI context. First, a climate data analysis with the aim of assessing the heat stress mitigation potential for different UGI scenarios in a typical neighborhood of Berlin, Germany was conducted. The second contribution was delivered in the form of CBAs with a focus on social benefits related to reduced morbidity and mortality rates as a result of UHI mitigation through UGI. While previous studies have addressed the UHI mitigation benefits of individual UGI measures, and stormwater management related UGI benefits on a neighborhood scale, this study proposes a method for analyzing complete UGI scenarios, including both ecosystem services and economic benefits related to reduced morbidity and mortality. The results of the climate data analysis indicate clear heat mitigation effects of UGI. The CBA results show that UHI mitigation through implementation of UGI measures at the neighborhood level is economically feasible. The methods for quantification of UGI project costs and benefits proposed in this study can be applied by urban planners in the objective evaluation of project proposals. Since UGI benefits create significant social value, the quantification thereof is essential for urban planning in providing a scientific foundation for the inclusion of UGI in UHI mitigation strategies. This is important as UGI installation and maintenance costs are high and delivering value to society can only be achieved through identification of the most suitable UGI measures and ensuring their effective placement.

Author Contributions: Conceptualization, D.J. and S.G.; methodology, J.E. and D.J.; formal analysis, J.E.; investigation, J.E.; data curation, D.J. and J.E.; writing—original draft preparation, J.E. and D.J.; writing—review and editing, S.G.; visualization, J.E.; supervision, S.G. All authors have read and agreed to the published version of the manuscript.

Funding: This research received no external funding.

Institutional Review Board Statement: Not applicable.

Informed Consent Statement: Not applicable.

Data Availability Statement: 3rd party data. Restrictions apply to the availability of these data. Data was obtained from the Institute of Meteorology and Climatology at Leibniz University Hannover and from GEO-NET Umweltconsulting GmbH and are available from the authors upon request with permission.

Conflicts of Interest: The authors declare no conflict of interest.

References

1. Oke, T.R. City size and the urban heat island. *Atmos. Environ.* **1973**, *7*, 769–779. [CrossRef]
2. Oke, T.R. The energetic basis of the urban heat island. *Q. J. R. Meteorol. Soc.* **1982**, *108*, 1–24. [CrossRef]
3. Mohajerani, A.; Bakaric, J.; Jeffrey-bailey, T. The urban heat island effect, its causes, and mitigation, with reference to the thermal properties of asphalt concrete. *J. Environ. Manag.* **2017**, *197*, 522–538. [CrossRef]
4. Fletcher, T.D.; Andrieu, H.; Hamel, P. Understanding, management and modelling of urban hydrology and its consequences for receiving waters: A state of the art. *Adv. Water Resour.* **2013**, *51*, 261–279. [CrossRef]
5. Grant, J.; Gallet, D. *The Value of Green Infrastructure a Guide to Recognizing Its Economic, Environmental and Social Benefits*; Center for Neighborhood Technology: Chicago, IL, USA, 2010; ISBN 1938-6478.
6. Herath, H.M.P.I.K.; Halwatura, R.U.; Jayasinghe, G.Y. Modeling a Tropical Urban Context with Green Walls and Green Roofs as an Urban Heat Island Adaptation Strategy. *Procedia Eng.* **2018**, *212*, 691–698. [CrossRef]
7. Santamouris, M.; Haddad, S.; Saliari, M.; Vasilakopoulou, K.; Synnefa, A.; Paolini, R.; Ulpiani, G.; Garshasbi, S.; Fiorito, F. On the energy impact of urban heat island in Sydney: Climate and energy potential of mitigation technologies. *Energy Build.* **2018**, *166*, 154–164. [CrossRef]
8. Buchin, O.; Hoelscher, M.-T.T.; Meier, F.; Nehls, T.; Ziegler, F. Evaluation of the health-risk reduction potential of countermeasures to urban heat islands. *Energy Build.* **2016**, *114*, 27–37. [CrossRef]
9. Dierkes, C.; Lucke, T.; Helmreich, B. General Technical Approvals for Decentralised Sustainable Urban Drainage Systems (SUDS)—The Current Situation in Germany. *Sustainability* **2015**, *7*, 3031–3051. [CrossRef]
10. Wicke, D.; Matzinger, A.; Rouault, P. *Relevanz Organischer Spurenstoffe im Regenwasserabfluss Berlins*; Kompetenzzentrum Wassetr: Berlin, Germany, 2015.
11. Fletcher, T.D.; Shuster, W.; Hunt, W.F.; Ashley, R.; Butler, D.; Arthur, S.; Trowsdale, S.; Barraud, S.; Semadeni-Davies, A.; Bertrand-Krajewski, J.-L.; et al. SUDS, LID, BMPs, WSUD and more—The evolution and application of terminology surrounding urban drainage. *Urban Water J.* **2015**, *12*, 525–542. [CrossRef]
12. Riechel, M.; Matzinger, A.; Pallasch, M.; Joswig, K.; Pawlowsky-Reusing, E.; Hinkelmann, R.; Rouault, P. Sustainable urban drainage systems in established city developments: Modelling the potential for CSO reduction and river impact mitigation. *J. Environ. Manag.* **2020**, *274*, 111207. [CrossRef]
13. Johnson, D.; Geisendorf, S. Are Neighborhood-level SUDS Worth it? An Assessment of the Economic Value of Sustainable Urban Drainage System Scenarios Using Cost-Benefit Analyses. *Ecol. Econ.* **2019**, *158*, 194–205. [CrossRef]
14. Hansen, R.; Pauleit, S. From multifunctionality to multiple ecosystem services? A conceptual framework for multifunctionality in green infrastructure planning for urban areas. *Ambio* **2014**, *43*, 516–519. [CrossRef]
15. Lähde, E.; Khadka, A.; Tahvonen, O.; Kokkonen, T. Can we really have it all?-Designing multifunctionality with sustainable urban drainage system elements. *Sustainability* **2019**, *11*, 1854. [CrossRef]
16. Gordon, B.L.; Quesnel, K.J.; Abs, R.; Ajami, N.K. A case-study based framework for assessing the multi-sector performance of green infrastructure. *J. Environ. Manag.* **2018**, *223*, 371–384. [CrossRef]
17. Caplan, J.S.; Galanti, R.C.; Olshevski, S.; Eisenman, S.W. Water relations of street trees in green infrastructure tree trench systems. *Urban For. Urban Green.* **2019**, *41*, 170–178. [CrossRef]
18. Saaroni, H.; Amorim, J.H.; Hiemstra, J.A.; Pearlmutter, D. Urban Green Infrastructure as a tool for urban heat mitigation: Survey of research methodologies and findings across different climatic regions. *Urban Clim.* **2018**, *24*, 94–110. [CrossRef]
19. Bodnaruk, E.W.; Kroll, C.N.; Yang, Y.; Hirabayashi, S.; Nowak, D.J.; Endreny, T.A. Where to plant urban trees? A spatially explicit methodology to explore ecosystem service tradeoffs. *Landsc. Urban Plan.* **2017**, *157*, 457–467. [CrossRef]
20. Larondelle, N.; Lauf, S. Balancing demand and supply of multiple urban ecosystem services on different spatial scales. *Ecosyst. Serv.* **2016**, *22*, 18–31. [CrossRef]
21. Zölch, T.; Maderspacher, J.; Wamsler, C.; Pauleit, S. Using green infrastructure for urban climate-proofing: An evaluation of heat mitigation measures at the micro-scale. *Urban For. Urban Green.* **2016**, *20*, 305–316. [CrossRef]
22. Carter, T.; Keeler, A. Life-cycle cost-benefit analysis of extensive vegetated roof systems. *J. Environ. Manag.* **2008**, *87*, 350–363. [CrossRef]
23. Joksimovic, D.; Alam, Z. Cost efficiency of Low Impact Development (LID) stormwater management practices. *Procedia Eng.* **2014**, *89*, 734–741. [CrossRef]
24. Liu, W.; Chen, W.; Feng, Q.; Peng, C.; Kang, P. Cost-benefit analysis of green infrastructures on community stormwater reduction and utilization: A case of Beijing, China. *Environ. Manag.* **2016**, *58*, 1015–1026. [CrossRef]
25. Perini, K.; Rosasco, P. Cost-benefit analysis for green façades and living wall systems. *Build. Environ.* **2013**, *70*, 110–121. [CrossRef]
26. Van Raalte, L.; Nolan, M.; Thakur, P.; Xue, S.; Parker, N. *Economic Assessment of the Urban Heat Island Effect*; AECOM Australia Pty Ltd.: Melbourne, Australian, 2012.
27. Venkataramanan, V.; Packman, A.I.; Peters, D.R.; Lopez, D.; Mccuskey, D.J.; Mcdonald, R.I.; Miller, W.M.; Young, S.L. A systematic review of the human health and social well-being outcomes of green infrastructure for stormwater and flood management. *J. Environ. Manag.* **2019**, *246*, 868–880. [CrossRef] [PubMed]
28. Hübler, M.; Klepper, G.; Peterson, S. Costs of climate change. The effects of rising temperatures on health and productivity in Germany. *Ecol. Econ.* **2008**, *68*, 381–393. [CrossRef]

29. Yu, S.; Xia, J.; Yan, Z.; Zhang, A.; Xia, Y.; Guan, D.; Han, J.; Wang, J.; Chen, L.; Liu, Y. Loss of work productivity in a warming world: Differences between developed and developing countries. *J. Clean. Prod.* **2019**, *208*, 1219–1225. [CrossRef]

30. Koppe, C.; Jendritzky, G. Die Auswirkungen von thermischen Belastungen auf die Mortalität. In *Warnsignale Klima: Gesundheitsrisiken. Gefahren für Pflanzen, Tiere und Menschen*; Lozan, J., Graßl, H., Jendritzky, G., Karbe, L., Reise, K., Eds.; Universität Hamburg, Institut f. Hydrobiologie, c/o Dr. J. Lozán: Hamburg, Germany, 2014.

31. Sahnoune, S.; Benhassine, N. Quantifying the Impact of Green-Roofs on Urban Heat Island Mitigation. *Int. J. Environ. Sci. Dev.* **2017**, *8*, 116–123. [CrossRef]

32. Scherer, D.; Fehrenbach, U.; Lakes, T.; Lauf, S.; Meier, F.; Schuster, C. Quantification of heat-Stress related mortality hazard, vulnerability and risk in Berlin, Germany. *DIE ERDE J. Geogr. Soc. Berlin* **2013**, *144*, 238–259. [CrossRef]

33. Hoelscher, M.-T.; Nehls, T.; Jänicke, B.; Wessolek, G. Quantifying cooling effects of facade greening: Shading, transpiration and insulation. *Energy Build.* **2016**, *114*, 283–290. [CrossRef]

34. Schubert, S.; Grossman-Clarke, S. The influence of green areas and roof albedos on air temperatures during extreme heat events in Berlin, Germany. *Meteorol. Z.* **2013**, *22*, 131–143. [CrossRef]

35. Von Tils, R. Effect of trees and greening of buildings on the indoor heating and cooling load—Microscale numerical experiment. *J. Heat Isl. Inst. Int.* **2017**, *12*, 35–39.

36. Simperler, L.; Ertl, T.; Matzinger, A. Spatial Compatibility of Implementing Nature-Based Solutions for Reducing Urban Heat Islands and Stormwater Pollution. *Sustainability* **2020**, *12*, 5967. [CrossRef]

37. Meyerhoff, J.; Dehnhardt, A. On the "non" use of environmental valuation estimates. In *Sustainability, Natural Capital and Nature Conservation*; Döring, R., Ed.; Metropolis: Marburg, Germany, 2009; pp. 143–166.

38. Gómez-Baggethun, E.; Ruiz-Pérez, M. Economic valuation and the commodification of ecosystem services. *Prog. Phys. Geogr.* **2011**, *35*, 613–628. [CrossRef]

39. Kallis, G.; Gómez-Baggethun, E.; Zografos, C. To value or not to value? That is not the question. *Ecol. Econ.* **2013**, *94*, 97–105. [CrossRef]

40. Rode, J.; Le Menestrel, M.; Cornelissen, G. Ecosystem service arguments enhance public support for environmental protection—But beware of the numbers! *Ecol. Econ.* **2017**, *141*, 213–221. [CrossRef]

41. Almansa, C.; Martínez-Paz, J.M. What weight should be assigned to future environmental impacts? A probabilistic cost benefit analysis using recent advances on discounting. *Sci. Total Environ.* **2011**, *409*, 1305–1314. [CrossRef]

42. Dasgupta, P. Discounting climate change. *J. Risk Uncertain.* **2008**, *37*, 141–169. [CrossRef]

43. Gowdy, J.; Rosser, J.B.; Roy, L. The evolution of hyperbolic discounting: Implications for truly social valuation of the future. *J. Econ. Behav. Organ.* **2013**, *90*, S94–S104. [CrossRef]

44. Matzinger, A.; Riechel, M.; Remy, C.; Schwarzmüller, H.; Rouault, P.; Schmidt, M.; Offermann, M.; Strehl, C.; Nickel, D.; Pallasch, M.; et al. *Zielorientierte Planung von Maßnahmen der Regenwasserbewirtschaftung–Ergebnisse des Projektes KURAS*; Konzepte fur Urbane Regenwasserbewirtschaftung und Abwassersysteme: Berlin, Germany, 2017.

45. DWA. *Leitlinien zur Durchführung Dynamischer Kostenvergleichsrechnungen (KVR-Leitlinien)*; 8. überarb; DWA Deutsche Vereinigung für Wasserwirtschaft, Abwasser und Abfall e. V.: Hennef, Germany, 2012; ISBN 9783941897557.

46. Baumgärtner, S.; Klein, A.M.; Thiel, D.; Winkler, K. Ramsey Discounting of Ecosystem Services. *Environ. Resour. Econ.* **2015**, *61*, 273–296. [CrossRef]

47. Drupp, M.A. Limits to Substitution Between Ecosystem Services and Manufactured Goods and Implications for Social Discounting. *Environ. Resour. Econ.* **2018**, *69*, 135–158. [CrossRef]

48. Attema, A.E.; Bleichrodt, H.; L'Haridon, O.; Peretti-Watel, P.; Seror, V. Discounting health and money: New evidence using a more robust method. *J. Risk Uncertain.* **2018**, *56*, 117–140. [CrossRef]

49. Frederick, S. Measuring Intergenerational Time Preference: Are future lives valued less? *J. Risk Uncertain.* **2003**, *26*, 39–53. [CrossRef]

50. Robinson, L.A.; Hammitt, J.K.; O'Keeffe, L. Valuing Mortality Risk Reductions in Global Benefit-Cost Analysis. *J. Benefit-Cost Anal.* **2019**, *10*, 15–50. [CrossRef] [PubMed]

51. Deutscher Wetterdienst. Erläuterungen und Kriterien zu Hitzewarnungen. Available online: http://www.wettergefahren.de/warnungen/hitzewarnungen.html (accessed on 5 April 2019).

52. R Core Team. *R: A Language and Environment for Statistical Computing*; R Foundation for Statistical Computing: Vienna, Austria. Available online: https://www.r-project.org/ (accessed on 17 January 2019).

53. Strehl, C.; Offermann, M.; Hein, A.; Heinzmann, B.; Thamsen, P.U.; Matzinger, A. *Schlussbericht des Forschungsvorhabens KURAS. IWW-Teilbericht: Ökonomische Effekte der Regenwasserbewirtschaftung am Beispiel Berlins*; KURAS: Berlin, Germany, 2017.

54. Eglitis-media. Sonnenaufgang und Sonnenuntergang in Deutschland. Available online: https://www.laenderdaten.info/Europa/Deutschland/sonnenuntergang.php (accessed on 16 March 2019).

55. Baccini, M.; Biggeri, A.; Accetta, G.; Kosatsky, T.; Katsouyanni, K.; Analitis, A.; Anderson, H.R.; Bisanti, L.; D'Iippoliti, D.; Danova, J.; et al. Heat effects on mortality in 15 European cities. *Epidemiology* **2008**, *19*, 711–719. [CrossRef]

56. Umweltatlas Berlin. Einwohnerdichte 2017 (Umweltatlas). Available online: https://fbinter.stadt-berlin.de/fb/index.jsp?loginkey=zoomStart&mapId=wmsk_06_06ewdichte2017@senstadt&bbox=389304,5822972,396530,5826388 (accessed on 4 March 2019).

57. Statistisches Bundesamt Deutschland. Durchschnittliches Sterbealter. Available online: https://www-genesis.destatis.de/genesis/online/link/tabelleErgebnis/12613-0007 (accessed on 19 March 2019).
58. Deutscher Wetterdienst. Deutschlandwetter im Jahr 2018. Available online: https://www.dwd.de/DE/presse/pressemitteilungen/DE/2018/20181228_deutschlandwetter_jahr2018.pdf?__blob=publicationFile&v=3 (accessed on 25 March 2019).
59. Deutscher Wetterdienst. Deutschlandwetter im Jahr 2017. Available online: https://www.dwd.de/DE/presse/pressemitteilungen/DE/2017/20171229_deutschlandwetter_jahr2017.pdf?__blob=publicationFile&v=2 (accessed on 25 March 2019).
60. Deutscher Wetterdienst. Deutschlandwetter im Jahr 2016. Available online: https://www.dwd.de/DE/presse/pressemitteilungen/DE/2016/20161229_deutschlandwetter_jahr2016.pdf?__blob=publicationFile&v=3 (accessed on 25 March 2019).
61. Deutscher Wetterdienst. Deutschlandwetter im Jahr 2015. Available online: https://www.dwd.de/DE/presse/pressemitteilungen/DE/2015/20151230_deutschlandwetter_jahr2015.pdf?__blob=publicationFile&v=2 (accessed on 25 March 2019).
62. Deutscher Wetterdienst. Deutschlandwetter im Jahr 2014. Available online: https://www.dwd.de/DE/presse/pressemitteilungen/DE/2014/20141230_Deutschlandwetter_Jahr_2014.pdf?__blob=publicationFile&v=7 (accessed on 25 March 2019).
63. Statistisches Bundesamt Deutschland. Aus dem Krankenhaus Entlassene Vollstationäre Patientinnen und Patienten (Einschließlich Sterbe- und Stundenfälle). Available online: https://www.destatis.de/DE/Themen/Gesellschaft-Umwelt/Gesundheit/Krankenhaeuser/Tabellen/entlassene-patienten-eckdaten.html (accessed on 21 March 2019).
64. Johnson, H.; Kovats, S.; Mcgregor, G.; JR, S.; Gibbs, M.; Walton, H. The impact of the 2003 heat wave on daily mortality in England and Wales and the use of rapid weekly mortality estimates. *Eurosurveillance* **2005**, *10*, 168–171. [CrossRef]
65. Tröltzsch, J.; Görlach, B.; Lückge, H.; Peter, M.; Sartorius, C. *Kosten und Nutzen von Anpassungsmaßnahmen an den Klimawandel: Analyse von 28 Anpasungmaßnahmen in Deutschland*; Umweltbundesamt: Dessau-Roßlau, Germany, 2012.
66. Statistisches Bundesamt Deutschland. Sterbefälle je 1000 Einwohner: Deutschland, Jahre. Available online: https://www-genesis.destatis.de/genesis/online/data;sid=C87037CAAF16C78B632D87B957981466.GO_1_4?operation=abruftabelleBearbeiten&levelindex=1&levelid=1553004933048&auswahloperation=abruftabelleAuspraegungAuswaehlen&auswahlverzeichnis=ordnungsstruktur&ausw (accessed on 19 March 2019).
67. Statistisches Bundesamt Deutschland. Bevölkerung: Bundesländer, Stichtag, Geschlecht, Altersjahre. Available online: https://www-genesis.destatis.de/genesis/online/data;sid=3A8959F91730FAB9CEAF0C0518849BB1.GO_1_4?operation=abruftabelleBearbeiten&levelindex=1&levelid=1552996107046&auswahloperation=abruftabelleAuspraegungAuswaehlen&auswahlverzeichnis=ordnungsstruktur&ausw (accessed on 19 March 2019).
68. Statistisches Bundesamt Deutschland. Durchschnittliche Lebenserwartung (Periodensterbetafel): Deutschland, Jahre, Geschlecht, Vollendetes Alter. Available online: https://www-genesis.destatis.de/genesis/online/logon?sequenz=tabelleErgebnis&selectionname=12621-0002&zeitscheiben=16&sachmerkmal=ALT577&sachschluessel=ALTVOLL000,ALTVOLL020,ALTVOLL040,ALTVOLL060,ALTVOLL065,ALTVOLL080 (accessed on 19 March 2019).
69. Viscusi, W.K.; Masterman, C.J. Income Elasticities and Global Values of a Statistical Life. *J. Benefit-Cost Anal.* **2017**, *8*, 226–250. [CrossRef]
70. OECD. Recommended Value of a Statistical Life numbers for policy analysis. In *Mortality Risk Valuation in Environment, Health and Transport Policies*; OECD Publishing: Paris, France, 2012; Volume 2, pp. 125–136.
71. Spengler, H. Kompensatorische Lohndifferenziale und der Wert eines statistischen Lebens in Deutschland. *Z. Arbeitsmarktforsch.* **2004**, *3*, 269–305.
72. Clark, C.; Adriaens, P.; Talbot, F.B. Green roof valuation: A probabilistic economic analysis of environmental benefits. *Environ. Sci. Technol.* **2008**, *42*, 2155–2161. [CrossRef]
73. MacMullan, E.; Reich, S.; Puttman, T.; Rodgers, K. Cost-Benefit Evaluation of Ecoroofs. In Proceedings of the Low Impact Development for Urban Ecosystem and Habitat Protection, Seattle, WA, USA, 16–19 November 2008; pp. 1–10.
74. Bianchini, F.; Hewage, K. Probabilistic social cost-benefit analysis for green roofs: A lifecycle approach. *Build. Environ.* **2012**, *58*, 152–162. [CrossRef]
75. Yang, J.; Yu, Q.; Gong, P. Quantifying air pollution removal by green roofs in Chicago. *Atmos. Environ.* **2008**, *42*, 7266–7273. [CrossRef]
76. Berliner Wasserbetriebe. Tarife für Trinkwasser, Schmutzwasser, Niederschlagswasser, Fäkalwasser und Fäkalschlamm vom 1. Januar 2016 bis 31. Dezember 2017. Available online: http://www.bwb.de/content/language1/html/204.php (accessed on 5 April 2018).
77. European Commission. *Guide to Cost-Benefit Analysis of Investment Projects. Economic Appraisal Tool for Cohesion Policy, 2014–2020*; European Commmission: Brussels, Belgium, 2014; ISBN 978-92-79-34796-2.
78. Groß, G.; von Tils, R. *Schlussbericht des Forschungsvorhabens KURAS -Teilbericht Leibniz Universität Hannover*; KURAS: Hannover, Germany, 2017.
79. Yazdanseta, A. Estimating the Cooling Power through Transpiration of Vining Green Walls in Various Climates. In *Symposium on Simulation for Architecture and Urban Design*; Turrin, M., Peters, B., O'Brien, W., Stouffs, R., Dogan, T., Eds.; Society for Modeling & Simulation International: Toronto, ON, Canada, 2017; pp. 235–242. ISBN 9781365888786.
80. Dang, T.N.; Van, D.Q.; Kusaka, H.; Seposo, X.T.; Honda, Y. Green Space and Deaths Attributable to the Urban Heat Island Effect in Ho Chi Minh City. *Am. J. Public Health* **2018**, *108*, S137–S143. [CrossRef]

81. Arzbach, V. Hitze-Notfälle: Die Schattenseiten des Sommers. Available online: https://ptaforum.pharmazeutische-zeitung.de/ausgabe-122018/die-schattenseiten-des-sommers/ (accessed on 1 February 2019).
82. Hu, K.; Guo, Y.; Hochrainer-stigler, S.; Liu, W.; See, L.; Yang, X.; Zhong, J.; Fei, F.; Chen, F.; Zhang, Y.; et al. Evidence for Urban—Rural Disparity in Temperature—Mortality Relationships in Province, Zhejiang. *Environ. Health Perspect.* **2019**, *127*, 1–11. [CrossRef] [PubMed]

MDPI
St. Alban-Anlage 66
4052 Basel
Switzerland
Tel. +41 61 683 77 34
Fax +41 61 302 89 18
www.mdpi.com

Sustainability Editorial Office
E-mail: sustainability@mdpi.com
www.mdpi.com/journal/sustainability